Losses in Water Distribution Networks

Losses in Water Distribution Networks

A Practitioner's Guide to Assessment, Monitoring and Control

Malcolm Farley and Stuart Trow

Publishing

Published by IWA Publishing, Alliance House, 12 Caxton Street, London SW1H 0QS, UK

Telephone: +44 (0) 20 7654 5500; Fax: +44 (0) 20 7654 5555; Email: publications@iwap.co.uk
Web: **www.iwapublishing.com**

First published 2003
© 2003 IWA Publishing

Printed by TJ International (Ltd), Padstow, Cornwall, UK
Copy-edited and typeset by HWA Text and Data Management, Tunbridge Wells, UK

British Library Cataloguing in Publication Data
A CIP catalogue record for this book is available from the British Library

ISBN: 1 900222 11 6

Contents

Preface

Water loss from distribution networks is a universal problem, requiring a management strategy which can be universally applied.

Developing such a strategy requires a diagnostic approach – first identify the problem and its causes, and then use appropriate tools to address the problem, and to reduce or remove it. To support the development of a universal strategy, the authors draw on first-hand knowledge of managing audits of water networks and water loss reduction projects for UK water companies, research and consultancy organisations, and international non-government organisations (NGOs).

The book will be of special interest to water network operators, consultants, and other industry practitioners who are about to embark on a water loss management programme, or who wish to review their existing procedures. The aim of the book is to allow water network operators to:

- understand the *significance and scale* of water loss components
- *quantify the volume* of losses by measurement or estimation
- become aware of the various techniques to *monitor and control* leakage
- select the appropriate *network management practices* and *equipment* to support the techniques
- *develop a strategy* for a particular network by considering the economic and other factors relevant to the local conditions and infrastructure

- *design a leakage management and control system* appropriate to local conditions and constraints
- be aware of the *operation and maintenance* and *training* requirements to sustain the policies and practices once in place

The book includes case study material from the authors' own experience and from that of colleagues in the international water industry.

The authors

MALCOLM FARLEY CEng FCIWEM

1982–1986: Research projects for the UK Water Research Centre, to develop the guidelines set out in *Leakage Control Policy and Practice* for leakage monitoring in zones within the network ('district metering'). This led to authorship of two WRc technical reports; *District Metering – System Design and Installation* and *District Metering – System Operation*

1987–1994: Project/contract manager for WRc leakage control projects in UK and Europe, and responsible for conducting training workshops for water engineers worldwide.

1991–1994: Steering Group member of, and Technical Secretary to, the UK National Leakage Initiative. Co-author of one of the *Managing Leakage* reports, *Report J – Techniques, Technology and Training*.

1994–2001: Independent consultant, specialising in training and consultancy for the international water industry. Short-term consultant for the World Bank, the World Health Organisation, the International Water Association, and UK Water Industry Research Ltd (UKWIR).

Publications include *A Manual of DMA Practice* (co-author), (UKWIR) and *Leakage Management and Control: A Training Manual* (WHO).

His specialism has been developing water loss strategies and training workshops in developing countries, some of which are used for case study examples in this book. He is now a principal of a newly formed consultancy – Aqua2 – specialising in water loss management, demand management, O&M and training projects worldwide.

Malcolm Farley can be contacted at mfarley@alvescot.demon.co.uk

STUART TROW BSc DMS CEng MICE MCIWEM

1979–1990: Management of leakage and water resource statistics, the development of district metering and pressure management, the management of leak detection teams, the operation of a central control telemetry facility, and the preparation of asset management plans (as required by OFWAT) while employed by the then Sunderland and South Shields Water Company (now part of Northumbrian Water).

1990–1994: Employed by North East Water (also now part of Northumbrian Water) as technical support manager responsible for all technical functions of water distribution. Member of the steering committee of the UK National Leakage Initiative. Co-author of one of the *Managing Leakage* reports, *Report C – Setting Economic Leakage Targets*.

1994–2000: Established and became managing director of Utility Technical Services Ltd. (now part of RPS Group plc), to provide consultancy, and contracted services to the greater majority of UK water suppliers on all aspects of water distribution leakage and losses. This included research work for UKWIR on various aspects of leakage economics and strategy, and assistance to several companies on setting leakage targets.

1998: Co-authored the Financial Times report *Managing Water Leakage – Economic and Technical Issues*.

2000–2002: Independent consultant, specialising in training and consultancy in the UK water industry on several aspects of water distribution management. Director of Fluid Controls (UK) Ltd., a product development company providing a range of flow and pressure management systems.

Stuart Trow can be contacted at stuarttrow@aol.com

Acknowledgements

The authors would like to acknowledge the high level of knowledge and expertise of their colleagues in the industry, several of whom have made inputs to the book:

Allan Lambert, for general guidance on topics of water balance, terminology and performance indicators, and for Chapter 3.
Mike Williamson, for inputs on topics of demand management and water conservation in Chapter 8.
Richard Taylor and Ecowater Systems, Auckland, for the water balance example in Chapter 2.
Pank Mistry and Sarina Shire Council, for Case Study 1.
Dewi Rogers and the municipalities of Gubbio and Brescia, Italy, for Case Study 2.
Alex Rizzo and the Water Works Corporation, Malta, for Case Study 3.

David Pearson and Lloyd Martin, for reviewing the book, and for providing invaluable comments.

The illustrations used in this book are taken from documents previously written by the authors and from other documents (used with the permission of the source organization), including reports published by the International Water Association, UK Water Industry Research Ltd., World Health Organization, and WRc plc.

Illustrations are reproduced with grateful thanks in particular to:

- Allan Lambert (for Figures 7.1, 7.2, 7.4 and 7.5, and for Table 7.1)
- Anglian Water (for Figure 8.1)
- The Department of the Environment (for Figure 8.2)
- David Pearson (for Figure 4.6)
- Ecowater Solutions (for Figure 2.3)
- International Water Association (for Figures 2.2, 3.2–3.6 and 4.3)
- UK Water Industry Research Ltd (for Figures 6.2–6.5, 7.8, 7.11, B1 and B2, and for Tables 5.2, 5.3 and 6.1)
- United Utilities (for Figure 7.7)
- WRc plc (for Figures 6.1, 7.3 and 7.9).

All other illustrations have been supplied by the authors themselves.

1

Introduction

1.1   UNDERSTANDING THE NETWORK

Globally, water demand is rising and resources are diminishing. Water loss from the pipe network, always the *bête noire* of the operations engineer, has long been a feature of operations management, even in countries with a well-developed infrastructure and good operating practices. However, it takes on a new dimension in developing countries, where a combination of poor infrastructure, poor sanitation, and intermittent supplies often poses a serious health risk. In water-scarce countries there is also a temptation to augment treated water with untreated in an attempt to satisfy demand.

Yet not all losses are the result of poor infrastructure and leaking pipes. 'Apparent' losses from the network, and excessive use or misuse of water, are often the result of local customs, combined with low tariff structures or inadequate metering policies. These losses, and the overuse of water, can often be reduced by the introduction of demand management and water conservation programmes alongside initiatives to tackle leakage and improve the pipe network. Together, these programmes form a strategy for restoring a potentially huge lost resource.

The key to developing a water loss strategy is to gain a better *understanding* of the reasons for losses and the *factors* which influence them. Then *techniques* and *procedures* can be developed, and tailored to the specific characteristics of the network and local influencing factors, to tackle each of the causes in order of priority. Whilst there is no quick and easy answer to reducing losses, much can be learned from the recent experiences of the UK water industry. Urged on by the regulators to develop 'more robust' procedures for analysing and controlling losses, industry practitioners now have in place a number of techniques for understanding, measuring, and monitoring losses within the distribution network. Most companies have also introduced programmes to encourage customers to use less water. These experiences are referred to throughout the book.

1.2   A STRATEGY FOR WATER LOSS

A diagnostic approach, followed by the implementation of solutions which are *practicable* and *achievable*, can be applied to any water company, anywhere in the world, to develop a water loss management strategy. However, practitioners and consultants working in developing countries will invariably face a slower pace, greater financial constraints, less developed infrastructure, lower levels of skills and technology, and political, cultural and social influences. These all have an influence on the scope for managing losses and demand, and affect the pace at which changes can be brought about. But the aim should always be to make some improvement to the current operational practice, by working together to transfer information, new skills, and motivation – to always leave something behind besides a report. The case studies referred to throughout the book exemplify the pitfalls, but also the successes, of introducing a strategy in developing countries as well as in those which have all the benefits of a well-developed infrastructure and sound operating policies.

Several definitive reports are quoted extensively throughout the book. The first – *Leakage Control, Policy and Practice* ('Report 26') [1], published in 1980, provided a starting point for UK water practitioners to examine their water losses and develop better techniques for controlling them. This was supported by two WRc reports which set out guidelines for district meter area (DMA) design and operation [2,3]. Ten years later, the UK National Leakage Initiative (1991–1994), and the report which resulted from the study [4], provided a model for the concept of understanding losses and developing solutions. This experience has been used throughout the book to demonstrate the development of a strategy for managing water loss, and the techniques for supporting it. During the last decade, much work has been done by the International Water Association and International Water Data Comparisons Ltd., to build on these documents, in order to provide a 'level playing field' for water suppliers worldwide to compare their performance against internationally recognised benchmarks.

The first step in developing a strategy is to ask some questions about the network characteristics and the operating practices, and then use the available tools and mechanisms to suggest appropriate solutions, which are used to formulate the strategy. Typical questions are:

- *How much* water is being lost?
- *Where* is it being lost from?
- *Why* is it being lost?
- *What* strategies can be introduced to *reduce* losses and *improve* performance?
- *How* can we *maintain* the strategy and *sustain* the achievements gained?

Figure 1.1 summarises the tasks required to address each question.

The book deals with these questions, tasks and solutions in each of the chapters. Each chapter is designed to 'stand alone', and for this reason there is some repetition of some of the concepts. For example, the range of technology for leakage management is described in outline in the chapter on strategy development, and is repeated in more detail in subsequent chapters. This is so that readers can 'dip' into each chapter according to their need for information and the level of advancement in the strategy process. For example, those practitioners with less-developed networks, and fewer resources, can proceed to the later chapters on upgrading the network, and designing leakage control and pressure control systems before embarking on the more detailed data analysis required for strategy development. At the end of the book are three examples of international water suppliers which have developed a water loss strategy to suit their particular needs. Case study examples are also presented in the text of some of the chapters – this is to provide continuity with the concepts explained in previous sections of the chapter. The book's content is structured as follows:

Chapter 1 introduces the concept of a strategy for water loss management.

Chapter 2 deals with the definition of water loss, the assessment of water loss and its components, and the techniques for measuring or estimating each component.

Chapter 3 introduces the IWA international standard as a means of comparing performance between countries.

Chapter 4 expands the theme of developing a strategy – using computer models to assess losses initially, and then using a model to formulate an appropriate strategy, setting economic targets and the means to achieve them.

QUESTION/SOLUTION	TASK
1. HOW MUCH WATER IS BEING LOST? – Measure components	WATER BALANCE – Improved estimation/measurement techniques – Meter calibration policy – Meter checks – Identify improvements to recording procedures
2. WHERE IS IT BEING LOST FROM? – Quantify leakage – Quantify apparent losses	NETWORK AUDIT – Leakage studies (reservoirs, transmission mains, distribution network) – Operational/customer investigations

3. WHY IS IT BEING LOST?
– Conduct network and operational audit
	REVIEW OF NETWORK OPERATING PRACTICES
	– Investigate:
	  • historical reasons
	  • poor practice
	  • quality management procedure
	  • poor materials/infrastructure
	  • local/political influences
	  • cultural/social/financial factors

4. HOW TO IMPROVE PERFORMANCE? – Upgrade the network – Design a strategy and action plans	UPGRADING AND STRATEGY DEVELOPMENT – Update records systems – Introduce zoning – Introduce leakage monitoring – Address causes of apparent losses – Initiate leak detection/repair policy – Design short/medium/long term action plans
5. HOW TO SUSTAIN PERFORMANCE? – Ensure sustainability with appropriate staffing and organisational structures	POLICY CHANGE, TRAINING AND O&M Training: – improve awareness – increase motivation – transfer skills – introduce best practice/technology O&M: – Community involvement – Water conservation/demand management programmes – Monitor action plan recommendations – Introduce O&M procedures

Figure 1.1 Tasks and tools for developing a strategy.

Chapter 5, acknowledging the wide range of infrastructure types, addresses how the network can be upgraded, and operational practices improved, to maximise the benefits of the strategy.

Chapter 6 describes the technologies for managing, monitoring,and controlling leakage from the network.

Chapter 7 describes the principles and practice of pressure management, one of the main planks of a water loss strategy.

Chapter 8 considers the policy changes required to tackle non-leakage or 'apparent' losses, and water regulatory and demand management issues

Chapter 9 gives direction on how the strategy can be sustained through operation and maintenance (O&M), and education and training programmes.

At the end of the book are three case studies, giving examples of how water company practitioners have tackled the strategy development issues and challenges raised in previous chapters, in a range of water supply undertakings across the world.

The appendices contain checklists and procedures which can be used to support some of the concepts outlined in the main body of the book.

1.3   REFERENCES

1   Technical Working Group on Waste of Water (1985 [1980]) *Leakage Control Policy and Practice*, Standing Technical Committee Report no. 26. Original publication London: Doe/NWC. Reprinted London: WAA/WRc.

2   Farley, M (1985) *District Metering Part 1: System Design and Installation*, WRc Report ER 180E. Swindon: WRc.

3   Farley, M and Merrifield, T (1987) *District Metering Part 2: System Operation*. WRc Report ER 210E. Swindon: WRc.

4   WSA/WCA Engineering and Operations Committee (1994) *Managing Leakage: UK Water Industry Managing Leakage* Reports A–J: Report A – *Summary Report*; Report B – *Reporting Comparative Leakage Performance*; Report C – *Setting Economic Leakage Targets*; Report D – *Estimating Unmeasured Water Delivered*; Report E – *Interpreting Measured Night Flows*; Report F - *Using Night Flow Data*; Report G – *Managing Water Pressure*; Report H – *Dealing With Customers' Leakage*; Report J – *Techniques, Technology and Training*.London: WRc/WSA/WCA.

2

Assessing losses

2.1   DEFINING WATER LOSS AND LEAKAGE

This section provides a general discussion on the definitions of water loss and leakage and its components, and the UK water industry approach to water balance calculation. Section 2.5 discusses the IWA international standard and provides a more detailed component analysis.

Water loss occurs in all distribution systems – only the volume of loss varies. This depends on the characteristics of the pipe network and other local factors, the water company's operational practice, and the level of technology and expertise applied to controlling it. The volume lost varies widely from country to country, and between regions of each country. The components of water loss, and their relative significance, also vary between countries. One of the cornerstones of a water loss strategy is therefore to understand the relative significance of each of the components, ensuring that each is measured or estimated as accurately as possible, so that priorities can be set via a series of action plans.

The expressions '*water loss*' and '*non-revenue water*' are now internationally accepted, and have replaced expressions such as 'unaccounted-for water' (UFW) which are less consistent and which make inter-country comparisons more difficult.

© 2003 IWA Publishing. *Losses in Water Distribution Networks* by Malcolm Farley and Stuart Trow. ISBN: 1 900222 11 6

Water losses = water produced – water billed or consumed

It is important to differentiate between water loss and leakage. The International Water Association has defined water loss as:

Water loss = 'real' losses + 'apparent' losses

The expression 'real losses' has replaced the expression 'physical losses'. 'Apparent' losses has replaced 'non-physical' losses, and 'management' losses. These expressions are described in detail in Chapter 3.

Real losses comprise *leakage* from pipes, joints and fittings, from leakage through service *reservoir floors and walls,* and from *reservoir overflows.* Real losses can be severe, and may go undetected for months or even years. The volume lost will depend largely on the characteristics of the pipe network and the leak detection and repair policy practised by the company, i.e.:

- the pressure in the network
- the frequency and typical flow rates of new leaks and bursts
- the proportions of new leaks which are 'reported'
- the 'awareness' time (how quickly the loss is noticed)
- the 'location' time (how quickly each new leak is located
- the repair time (how quickly it is repaired or shut off)
- the level of 'background' leakage (undetectable small leaks)

These influences were acknowledged in the research programme undertaken by the UK National Leakage Initiative [1]. The vast differences in volume lost from leaks in three different parts of the distribution network are illustrated in Figure 4.11 in Section 4.7.1. Figure 2.1 illustrates a burst distribution main, which would normally be repaired or shut off immediately, and a service pipe leak, which, if underground, could go undetected for months.

Figure 2.1 Leaks on a main and a service pipe.

Leakage is usually the major component of water loss in *developed* countries, but this is not always the case in *developing* or *partially developed* countries, where illegal connections, meter error, or accounting errors are often more significant. This was acknowledged in a survey commissioned by the International Water Supply Association in 1991 [2]. Water companies from 14 countries volunteered data on annual losses, some as average national figures, others as individual water companies. Almost without exception, leakage is the major component. Exceptions were Barcelona, in Spain, and Singapore, where meter under-registration was the major component.

2.2   COMPARISONS OF WATER LOSSES

The International Water Association (IWA) has standardised the methods of defining and expressing total water losses [3], and supports the UK practice of always calculating and initially expressing components of water losses as a volume per year or per day, before any comparisons are attempted.

Recent work by Task Forces of the IWA has now substantially resolved long-standing problems associated with making meaningful comparisons of water loss between systems and countries with diverse characteristics. These recommendations are presented and discussed in Chapter 3.

2.3   EXPERIENCE OF THE UK WATER INDUSTRY

Whilst the majority of losses in the UK water industry are mostly 'real' losses, i.e. bursts from fractured pipes and background leakage in the network, meter error can be a significant contributor to 'apparent' losses in some companies. All of the UK water undertakings practice active water loss management, but it is the water companies of England and Wales which have pioneered the mechanisms and technologies for gaining a better understanding and control of water losses. This has been under stringent regulation by the industry regulators – the Office of Water Services (OFWAT) and the Environment Agency (EA). These mechanisms are now being applied in Scotland and Northern Ireland.

The water companies of England and Wales comprise 10 large 'water service' companies, which supply water and dispose of sewage, and 16 smaller companies which supply water only to discrete supply areas. Since 1989 the water companies of England and Wales have been in the private sector, and are regulated under the 1991 Water Act. These companies ensure that 99% of the population of England and Wales is served with a piped supply of water. A total of 15 260 Ml was supplied daily to 50 million people (21 million households) via 300 000 km of water mains in 2000/01.

The two main water industry regulators, OFWAT and the EA, have put increasing pressure on water companies to reduce losses in their networks. In particular, OFWAT introduced a stringent reporting procedure, where each water company is required to submit a detailed annual report on the volume of water supplied, consumed, and lost in each component part of the network – using a *water balance* cross-checked by *night flow analyses*. The basic terminology was developed during the UK National Leakage Initiative [1] and is now used by OFWAT to make *inter-company comparisons,* using each company's performance as a measure of its *efficiency*. OFWAT sets each company a mandatory annual leakage *target in Ml/ day*, which it must achieve to avoid sanctions. In recent years such drivers have led to water companies making great strides in reducing losses.

OFWAT analyses companies' water balance components and currently *reports* real losses (leakage) as:

• distribution losses, expressed in Ml/day, litres/property/day, and $m^3/km/day$
• losses on private supply pipes, expressed in Ml/day and litres/property/day

Total leakage in England and Wales fell from 5 112 Ml/day in 1994/1995 to 3 306 Ml/day in 1999/2000 (2 431 Ml/day distribution losses + 875 Ml/day supply pipe losses), a reduction of 35%. All but two companies met their targets for 1999/ 2000, an industry reduction in leakage of 245 Ml/d (7%) from the previous year. A further 9% reduction would have occurred if all companies had achieved their targets for 2000/2001, but only 2% reduction was achieved that year with leakage of 3 245 Ml/day. Leakage rose by 5% to 3 410 Ml/day in 2001/2002. In the 2000/2001 OFWAT report [4] the regulator notes that 19 of the 22 companies met their targets, and this emphasises the contribution made to increased leakage by the other three.

Since the National Leakage Initiative published its findings in 1994 [1] research and development has continued under the auspices of UK Water Industry Research Ltd (UKWIR). This organisation commissions and contracts research and development (R&D) on behalf of its member companies, and leakage features highly among its priorities. Work has continued on methodologies and mechanisms for accurately estimating and measuring losses, for calculating economic leakage targets, and technologies for detecting leakage in 'difficult' situations, such as transmission mains.

For many years the 'Cinderella' of the UK water industry, water loss management in recent years has become one of the major operational tasks of the distribution network. This has resulted from a combination of privatisation and regulation, making companies increasingly accountable to customers, shareholders and regulators. The substantial reduction of water losses since 1994 is largely a result of meeting these pressures, and the stringent reporting requirements. The *regulatory reporting* procedure for England and Wales is probably the most stringent in the world. It is possible that countries which do not have such a procedure are seriously under-estimating losses. The water companies of England and Wales now have a detailed *understanding* of the components of losses and the technology for measuring them. All water companies are now 'transparent' in calculating and publicising comparable independently audited figures for losses.

2.4   WATER BALANCE – UK METHODOLOGY

Water loss can be determined by conducting a *water balance*. This is based on the measurement or estimation of water produced, imported, exported, consumed or lost – the calculation should balance. Most water companies are able to provide estimates of production, imports, exports and consumption, but less able to estimate the other components. The water balance calculation provides a guide to how much is lost as leakage from the network ('real' losses), and how much is due to 'apparent' or non-physical losses.

In 1994 the report *Managing Leakage* [1], produced by the UK National Leakage Initiative, defined the terminology for the components of 'distribution input' (water produced) which has become the UK national standard, and the basis for the water balance calculation required by OFWAT:

Distribution input (DI)

      = Water produced(WP) + water imported(WI) – water exported (WE)

      = water delivered (WD) + distribution losses (DL) + distribution operational use
      (DOU)

'Water delivered' is the volume reaching the customer's boundary, and includes measured and unmeasured use, unmeasured supply pipe losses, and minor losses from legal and illegal use from hydrants. 'Distribution losses' includes all losses of potable water between the treatment works and the highway boundary. 'Supply pipe losses' is leakage from customers' private pipes after the point of delivery. As not all UK households have revenue meters, customer consumption, including customer supply pipe leakage, has to be estimated. Water companies have been encouraged by the regulators to implement more 'robust' mechanisms for estimating the unmeasured components of water delivered to give greater accuracy to the figures for supply pipe losses.

2.5   THE IWA INTERNATIONAL STANDARD

Because of the wide diversity of formats and definitions used for water balance calculations internationally (often within the same country), there has been an urgent

System input volume (corrected for known errors)	Authorised consumption	Billed authorised consumption	Billed metered consumption (including water exported)	Revenue water
			Billed unmetered consumption	
		Unbilled authorised consumption	Unbilled metered consumption	Non-revenue water (NRW)
			Unbilled unmetered consumption	
	Water losses	Apparent losses	Unauthorised consumption	
			Customer metering inaccuracies	
		Real losses	Leakage on transmission and/or distribution mains	
			Leakage and overflows at utility's storage tanks	
			Leakage on service connections up to point of customer metering	

Figure 2.2 IWA standard international water balance and terminology.

need for a common international terminology. Drawing on the best practice from many countries, IWA Task Forces on Water Losses and Performance Indicators recently produced an international 'best practice' standard approach for water balance calculations (Figure 2.2), with definitions of all terms involved.

Abbreviated definitions of principal components of the IWA water balance are as follows:

- System input volume is the annual volume input to that part of the water supply system.
- Authorised consumption is the annual volume of metered and/or non-metered water taken by registered customers, the water supplier and others who are implicitly or explicitly authorised to do so. It includes water exported, and leaks and overflows after the point of customer metering.
- Non-revenue water (NRW) is the difference between system input volume and billed authorised consumption. NRW consists of:
  - unbilled authorised consumption (usually a minor component of the water balance)
  - water losses
- Water losses are the difference between system input volume and authorised consumption, and consists of apparent losses and real losses.

- Apparent losses consist of unauthorised consumption and all types of metering inaccuracies.
- Real losses are the annual volumes lost through all types of leaks, bursts and overflows on mains, service reservoirs and service connections, up to the point of customer metering.

The components of the water balance should always be calculated as volumes before any attempt is made to calculate performance indicators. The separation of non-revenue water into components – unbilled authorised consumption, apparent losses and real losses – should always be attempted.

Widespread international use of the IWA standard water balance is being encouraged, as the first step in calculating the IWA 'best practice' performance indicators. The IWA standard water balance is gaining rapid acceptance, and has already been adopted or promoted (with minor variations) by

- national organisations in Australia [5], Germany [6], Malta [7], New Zealand [8], South Africa [9] and the USA
- individual utilities or consultants in the above countries and Brazil, Canada, Malaysia, the Nordic countries and the UK

It is usually relatively easy to re-allocate the components of any 'national' or 'local' water balance into the IWA standard approach, before calculating the annual volumes of losses and the 'best practice' performance indicators. If a national standard is being proposed or reviewed, as in Germany [6], the IWA standard approach should be the first logical choice.

All components of the water balance, and the performance indicators derived from it, are subject to errors in input data. Accordingly, recent applications of the methodology as described in [8, 10], increasingly use software with provision for entering 95% confidence limits for all data entry items, and automatic calculation of 95% confidence limits for NRW components and performance indicators. An example water balance calculation is shown in section 2.6. The following sections define non-revenue water and its components – unbilled authorised consumption, apparent losses and real losses.

2.5.1   Non-revenue water

The volume of non-revenue water is calculated by deducting the volumes of components of billed authorised consumption from the system input volume. The procedure for carrying out the remainder of the water balance calculation, to assess components of non-revenue water, consists of assessing the unbilled authorised consumption and apparent losses, and deducting these from the NRW volume to obtain the real losses.

Because of widely varying interpretations of the term 'unaccounted-for water' (UFW) worldwide, the IWA Task Forces do not recommend use of this term. If the term UFW is used at all, it should be defined and calculated in the same way as 'non-revenue water' in Figure 2.2.

2.5.2   Unbilled authorised consumption

Authorised consumption, in the IWA terminology, includes items such as fire fighting and training, flushing of mains and sewers, cleaning of suppliers' storage tanks, filling of water tankers, water taken from hydrants, street cleaning, watering of municipal gardens, public fountains, frost protection, building water etc. These may be billed or unbilled, metered or unmetered, according to local practice.

Unbilled authorised consumption should normally be only a small component of the water balance (less than 1% of system input volume). Wherever feasible, such volumes should be metered. In other situations, simple methods of documenting and estimating such uses often show that volumes of unbilled authorised consumption are unnecessarily high, and can be managed down to smaller annual volumes without influencing operational efficiency or customer service standards.

2.5.3   Apparent losses

Apparent losses consist of unauthorised consumption (theft and illegal use) and metering errors. Calculations of these volumes are preferably based on structured sampling tests, or estimated by a robust local procedure (which should be defined for audit purposes).

When quoted as a percentage of system input volume, apparent losses can range from almost zero to 10% for direct pressure systems, or even more for systems with customer storage tanks. Accordingly, each utility should attempt to assess and manage the components of apparent losses for its own system(s).

2.5.4   Unauthorised consumption

Unauthorised consumption occurs to a greater or lesser extent in most systems worldwide, but in reasonably well managed systems it should not exceed 1% of system input volume – the England and Wales estimate is 0.36% of system input volume.

This component of apparent losses, is generally associated with misuse of fire hydrants and fire service connections, and illegal connections.

Sample metering or pressure management of normally unmetered fire services may identify such misuse.

Checking for possible illegal connections often commences with identification of customers with unusually low consumption.

In the USA, the M36 Audit procedure recommends that unauthorised consumption is best dealt with through 'good billing procedures'.

2.5.5   Assessment of metering errors

It is preferable to correct any known errors to the system input volume at the start of the Water Balance calculation. This not only reinforces the absolute necessity for regular checks on the accuracy of the system input meters, but also seeks to ensure that the only metering errors in Apparent Losses will be customer metering errors, making the calculated volumes easier to interpret.

Customer metering errors include:

- random errors due to accounting procedures – differences between dates of source meter readings and customer meter readings, misread meters, incorrect estimates for stopped meters, adjustments to original meter readings, improper calculations, computer programming errors, etc.
- systematic errors due to under-registration or over-registration of customer meters

Systematic under/over registration of customer meters depends on many factors – for example, the type and class of meter, the method of installation, the water quality, the continuity of supply, the average working life of meters and the presence (or absence) of storage tanks on customers' properties.

Management of customer metering errors

Many computer-based billing systems are, unfortunately, not designed for efficient retrieval of technical data for water losses studies and calculations. Improved liaison between the billing and operational sections of the utility can minimise such problems. If there are spare 'fields' in the billing system, the use of global positioning system (GPS) references for each customer meter location may, in some situations, offer a way of assigning individual meters to individual sectors. Geographical information systems (GIS) technology also provides the capability to link every customer meter with a service connection.

The selection of customer meter types and classes (A to D) may be limited by water quality considerations, as well as technical and economic considerations. Economic replacement policies for residential meters based on selective testing programmes generally indicate changeover periods between 5 and 10 years. Incorrectly sized commercial meters can result in significant under-registration of consumption, and checks can be made to identify if there are more appropriate meters for individual situations (by occasional monitoring of the actual frequency and range of consumption rates).

When samples of customer meters are tested for accuracy, it is normal to quote the error as a percentage of the recorded metered consumption. Where customers are served by way of roof tanks, the probability of customer meter under-registration is greatly increased, because of the tendency for a greater part of the consumption to pass through the meter at rates less than the minimum flowrate, Q_{min}, specified for the meter.

2.5.6   Real losses

Although a water balance should always be attempted, there are disadvantages in relying only on water balance for assessment of real losses:

- The accumulated errors from the other components will be associated with the estimate of real losses.
- A water balance normally covers a 12-month retrospective period, so it has limited value as an 'early warning' system for identifying new unreported leaks and bursts, and initiating active leakage control to limit their duration.
- The water balance gives no indication of the individual components of real losses, or how they are influenced by utility policies.

For these reasons, real losses should preferably also be assessed by additional methods, namely:

- component analysis of real losses
- analysis of night flows

2.5.7   Component analysis of real losses

The general principle of assessing some components of real losses from repair statistics is well known. Annual numbers of repairs are assumed to represent the annual number of new leaks and bursts; these are then classified into different categories, with different typical flow rates. If average duration of each category of leak or burst is logically assessed, based on utility policies, then the annual volume lost from different categories can be assessed. In 1993, an internationally applicable overview concept known as 'Background and Bursts Estimates' (BABE) was developed from this basic building block, for calculating components of real losses based on the parameters which influence them [11].

In BABE analyses, components of real losses are considered to consist of:

Table 2.1 Parameters used for component analysis of annual real losses.

Component of infrastructure	Background undetectable) leakage	Reported bursts and overflows	Unreported bursts and overflows
Mains	Length Pressure Min loss rate/km*	Number/year Pressure Average flow rate* Average duration	Number/year Pressure Average flow rate* Average duration
Service reservoirs	Leakage through structure, percentage of capacity/day	No. of reported overflows/yr Average flow rate Average duration	No. of unreported overflows/year Average flow rate Average duration
Service connections, main to property boundary	Number of service connections Pressure Min loss rate/conn*	Number/year Pressure Average flow rate* Average duration	Number/year Pressure Average flow rate* Average duration
Service connections after property boundary	Length Pressure Min loss rate/km*	Number/year Pressure Average flow rate* Average duration	Number/year Pressure Average flow rate* Average duration

* at some standard pressure, later corrected for actual pressure using the FAVAD concept (see Chapter 7)

- background leakage at joints and fittings, flow rates too low for sonic detection if non-visible
- reported leaks and bursts – typically high flow rates but short duration
- unreported leaks and bursts – moderate flow rates, average duration depends on method of active leakage control

By considering average duration of detectable leaks and bursts to consist of three components – awareness, location and repair time – these concepts can be used to model any utility policies and standards of service. Typical burst flow rates are specified at a standard pressure, and adjusted to actual pressure using appropriate assumptions for pressure : leakage relationship. The typical parameters which would be assumed to influence components of annual real losses in different parts of the infrastructure are shown in Table 2.1.

The concept of BABE is described in more detail in section 4.7.

2.5.8   Assessing annual real losses from continuous night flows

Night flows measured in moderately-sized sectors (up to around 3000 service connections) are extremely useful for identifying the presence of existing unreported leaks and bursts, and the occurrence of new ones.

However, continuous night flows can also be used for assessing annual average real losses. Night flows in individual sectors must be measured continuously throughout the year. Customer night consumption must be assessed and deducted, and the average night leakage (in m^3/hour) must be multiplied by an 'hour-day factor' which depends upon the 24-hour variation of average pressure in the sectors.

The Economic Regulator (OFWAT) for England and Wales requires water companies to calculate annual real losses by this 'bottom-up' method [4], as well as using the 'top-down' water balance. Both methodologies are used by Malta Water Works Corporation (see Case Study 3).

2.5.9   Where do the largest components of real losses occur?

Analyses of components of real losses assist in identifying where the largest components typically occur in any individual system, and how these components may be influenced by utility policies.

The large number of joints and fittings on service connections between the main and the street/property boundary result in a relatively high value for background leakage in this part of the infrastructure. Also, studies of new break frequencies on mains and services in Germany [12] and Portugal [13] show that the frequency of new bursts, per kilometre of pipe, is several times higher on service connections than on mains. Although average burst flow rates are higher for mains than for service connections, when typical frequencies of reported and unreported bursts, and average durations of different types of bursts, are taken into account, it is evident that in most systems the largest volume of annual real losses generally occurs on service connections. This accords with most operational experience world-wide.

There will of course be some systems where the greatest proportion of real losses will be associated with the mains, rather than the service connections. The Water Losses Task Force calculated that, in well-managed systems, this 'break-point' occurs when the density of connections is around 20 per km of mains, and this figure has implications for the choice of the most appropriate performance indicators.

There are many different situations regarding ownership and maintenance responsibility of service connections, which can have a major influence on the annual volume of real losses. For example, in Finland, Japan, Norway, and parts of the USA, the whole of the service connection, from the main to the customer meter, is the customers' responsibility, and customer meters can be located anywhere from the street/property boundary to 30 metres or more after the boundary.

In many systems, the customer meter is located close to the street/property boundary, and the service pipe between the main and the customer meter is owned and maintained by the water utility. Where the customer meter is located some distance after the street/property boundary, the leakage on the private pipes between the street/property boundary and the customer meter becomes an additional component of calculated real losses. Detailed performance comparisons therefore need to allow for customer meter location relative to the street/property boundary. These are discussed in Chapter 3.

2.6.   AN EXAMPLE WATER BALANCE CALCULATION

Figure 2.3 shows an example of a water balance calculation with 95% confidence limits, for a fully metered system in New Zealand. Ecowater Solutions is a water distribution utility serving the north-west part of Auckland, and the water balance data is taken from the 'Benchloss' software [8] which has recently been completed for the New Zealand Water and Waste Association, using the IWA 'best practice' principles. This software also calculates the IWA recommended performance indicators.

Ecowater has an excellent record in leakage management. The calculated infrastructure leakage index (ILI) for 2000/01 is fractionally below 1.0, and extensive pressure management has been introduced over the past five years.

The introduction of 95% confidence limits to the calculation provides a particularly interesting perspective. Despite the 95% confidence limits for the metered input and authorised consumption volumes being 2% or less, at these low leakage levels the 95% confidence limits are of the order of:

- ±23% for non-revenue water and water losses
- ±50% for apparent losses
- ±28% for real losses

These figures clearly demonstrate that NRW and NRW components calculated from water balance are not exact figures, even in fully metered systems. Checking based on component analysis (used by Ecowater for evaluating alternative active leakage control strategies) and/or night flow analysis can help to reduce the uncertainty – although of course both these methods are also subject to 95% confidence limits.

System input volume 14 321m³ × 10³ ±2.0%	Authorised consumption 12 791 m³×10³ ±1.6%	Billed authorised consumption 12 747 m³×10³ ±1.6%	Billed metered consumption (including water exported) 12 747 m³ × 10³ ± 1.6%	Revenue water 12 747 m³ × 10³ ±1.6%
			Billed unmetered consumption Zero	
		Unbilled authorised consumption 44 m³ × 10³ ±29.4%	Unbilled metered consumption Zero	Non-revenue water (NRW) 1 574 m³ × 10³ ±22.6%
			Unbilled unmetered consumption 44 m³ × 10³ ±29.4%	
	Water losses 1530 m³ × 10³ ±23.2%	Apparent losses 232 m³ × 10³ ±52.2%	Unauthorised consumption 3 m³ × 10³ ±80%	
			Customer metering inaccuracies 229 m³ × 10³ ±52.7%	
		Real losses 1 298 m³× 10³ ±28.9%	Real losses 1 298 m³ × 10³ ±28.9%	

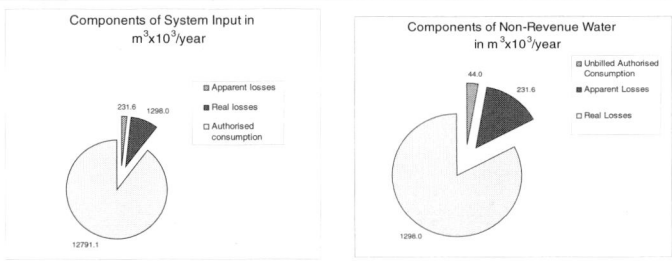

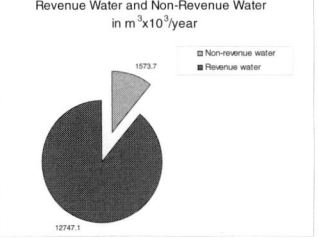

Figure 2.3 Water balance for Ecowater Solutions system, July 2000 to June 2001.

2.7   MEASURING OR ESTIMATING THE COMPONENTS OF WATER BALANCE

The components of water balance shown in Figure 2.2 can be measured or estimated using a number of techniques – the most appropriate ones for the system under investigation should be selected. Ideally, all components of the water balance should be quantified over the same designated period, and expressed in volumetric terms before attempting performance comparisons.

2.7.1   System input volume

This step identifies all water sources and quantifies water supplied. Measurement is by one or more of the following:

*   existing production (bulk) meters, after checking accuracy
    – summation of zone flows where there are existing zone meters
*   reservoir drop tests
*   checking pump curves
*   insertion meters at points where there are no meters, e.g. upstream or downstream of a treatment works (if upstream, losses during the treatment process should be assessed and subtracted)
*   over-estimation of system input is from
    – inadequate or no measurement facility
    – inadequate calibration programme for supply meters
    – systematic errors due to inadequate knowledge of the network configuration

Authorised consumption

BILLED AUTHORISED CONSUMPTION

BILLED METERED CONSUMPTION
*   Billing records are used to quantify measured outputs from the system.
*   Identify all non-household customers.
*   Identify all household customers.
*   Convert billing data over a monthly or quarterly billing period to an average daily flow in Ml/day.

BILLED UNMETERED CONSUMPTION.
*   This step identifies unmetered households and other authorised unmeasured consumption.
*   Estimate consumption by unmetered households. Monitor a sample of households by meter, or estimate per capita consumption.

UNBILLED AUTHORISED CONSUMPTION

- Identify and estimate measured and unmeasured consumption by authorised users, e.g. municipal buildings, parks, fire services, tankers
- Identify and estimate unmeasured supplies to peri-urban areas (slum areas, squatters etc.)
- Identify and estimate water used by the company for operational purposes, e.g. mains cleaning and flushing

Water losses

Adding volumes for authorised consumption, and subtract them from system input, will give water losses. These constitute real and apparent losses

REAL LOSSES

Real losses are:

- leakage from reservoirs
- overflows from reservoirs
- leakage from transmission mains
- leakage from the distribution network – from the company's mains, service connections, and fittings

RESERVOIR LEAKAGE

This can be measured by reservoir drop test. Inlet and outlet valves are closed and the drop in water level is measured over time. If the reservoir has compartments, each can be monitored in turn. Drop tests are usually carried out at night to minimise disruption to supply.

RESERVOIR OVERFLOWS

Overflows from reservoirs are caused by ineffective inlet controls such as faulty float valves. These should be inspected when the reservoir is at top water level to identify those which are passing. The volume passing is measured over the period when the reservoir is at top water line.

LEAKAGE FROM TRANSMISSION MAINS

Use insertion meters or clamp-on ultrasonic meters at each end of the main to calculate change in volume flowrate. Alternatively include transmission mains in distribution system measurement

LEAKAGE IN THE DISTRIBUTION SYSTEM

Supply zone measurement. This method measures night flow into a supply zone and includes transmission main leakage. Use a reservoir bulk meter to monitor flows at night (0200 – 0400). Subtract measured night use by large customers and estimated night use and leakage by households and non-households. If there are no bulk meters, it may be possible to use another form of reservoir drop test. Close the inlet valve to the reservoir and measure the drop in level over the same period and subtracting the same values for customer consumption. This calculation will also include reservoir leakage and transmission main leakage.

District Meter Areas (DMAs). DMAs are small zones of usually 500–3000 households within a supply zone. Each DMA is a discrete zone, with a defined and permanent boundary. Flows into (and out of) each DMA are monitored by a flowmeter. Night flow measurements are used to calculate leakage in the distribution system (distribution mains, service connections and fittings), after subtracting customer night use and customer leakage. To convert night leakage rates to daily average leakage, the night leakage rates must then be multiplied by an 'hour-day factor' [14] which allows for the variation of average pressure (and thus leakage rates) from midnight to midnight.

APPARENT LOSSES

Apparent losses are those which are not due to leakage. They are often the result of influences on the company which are usually beyond the scope of day-to-day operational practice. Apparent losses can be influenced by social and cultural factors, political influences, and financial factors, and often require institutional and organisational changes. As such they are more difficult to address, and are usually part of a medium- to long-term action plan within a water loss strategy. The plan may require changes to be made to:

- community lifestyles and education, e.g. the introduction of demand management and water conservation programmes
- cost recovery policy, e.g. the revision of tariff structures and charging policy and customer metering policy
- revenue collection, e.g. a review of staffing policy and training
- policy on illegal connections, particularly in peri-urban and slum areas, involving political decisions, and the human right to water as a necessity for life
- metering policies to improve accuracy of measurement and estimation, such as meter installation, calibration, and repair programmes

UNAUTHORISED CONSUMPTION

This step identifies and estimates illegal supplies and theft, caused by:

- illegal connections
- misuse of fire hydrants and fire services
- vandalised meters, bypassed meters
- bribery and corruption of meter readers.

Estimate the number of illegal connections from historic records and anecdotal evidence of inspectors, or by a house-to-house survey of a sample zone (check that each connection has a billing reference)

CUSTOMER METERING INACCURACIES

Under-estimation of consumption is caused by:

- under-registration of customer meters (particularly where customers are supplied through roof tanks)
- poor quality, inaccurate meters
- stopped meters
- inadequate meter maintenance/replacement policy
- inadequate meter reading policy
- under-estimation of free supplies or operational use
  - Estimate under-registration by taking a sample of metered households and installing a check meter, usually a Class D inferential meter, downstream of the customer meter. Percentage under-registration can be calculated. Results, if significant, can guide a meter replacement programme
  - Estimate the number of broken or by-passed meters
  - Use per capita consumption estimates to calculate volume used

The water balance calculation at an early stage of strategy development will help to find the significance, scale, and causes of the apparent losses, and will help to prioritise action plans to correct them.

Wasteful use

Although not a component of water balance, wasteful or excessive use is often a feature of the same constraints and influences as those for apparent losses, and should be addressed by parallel policy changes. Examples are:

- inadequate customer metering policy
- inappropriate charging policy (flat rate tariff, subsidised supplies)
- cultural and social traditions
- inadequate community education policy

2.8   TECHNIQUES FOR REVIEWING THE NETWORK OPERATING PRACTICES

This section addresses the question 'why is water being lost'? It reflects the company's management of its network, and can be answered by carrying out an appraisal of the physical characteristics of the network and the current operational practice. The review usually reveals the good practices as well as the problems caused by poor infrastructure and bad management practice. Chapter 5 outlines a programme of improvements to upgrade the network so that water loss can be assessed, monitored, and controlled more easily.

The appraisal should assess:

- particular country or regional characteristics, influencing factors, components of water loss
- the condition of the network
- current practice and methodologies used for operating and managing the network, including the facilities for monitoring flows, pressures and reservoir stocks
- the level of technology for monitoring and detecting leakage
- staff skills and capabilities

Particular tasks should include:

a)   *Discussions with senior staff* – i.e. directors and senior managers on current management practice, perceptions, financial and political constraints and influences, and future planning.
b)   *Discussions with operational staff* on system features and practice, including:
  - physical data (population, demands, topography, supply arrangements, mains length, number of service connections, customer meter location, average pressure)
  - drawings and records, billing data
  - measurements or estimates of system input volumes
  - estimates of authorised and unauthorised consumption estimates of non-revenue water components and performance indicators based on the IWA approach, with confidence limits
  - current practice (staffing structure, staff numbers and skills)
  - techniques and equipment
  - repair programme
  - economic data (cost of water etc.)
c)   *Field visits* – to appraise current practice and skills
d)   *Selection of a suitable pilot area* – for a future project to demonstrate techniques and equipment, gather results and show benefits, and to train staff.

2.9   TECHNIQUES FOR QUANTIFYING LEAKAGE

This section explains in more detail the techniques for measuring leakage in reservoirs, transmission mains, and the distribution network. Although leakage in reservoirs and transmission mains can be significant, the majority of leakage occurs in the mains and service pipes within the distribution network. The procedure adopted for leakage measurement depends on the supply arrangements and the design characteristics of the supply and distribution system. The main factors to consider are whether:

a)   there is 24-hour or intermittent supply
b)   supply arrangements can be temporarily changed
c)   supply is to discrete zones with designated boundaries, or to the total distribution area

If there is 24-hour supply, or the system will allow temporary rearrangement to provide 24-hour supply during the test period, leakage measurement can be made using the total night flow method. If these arrangements cannot be made, then leakage must be estimated using the total quantity method. The extent to which each of the components of the distribution system can be measured depends on the design of the supply and distribution system. For example, in a system which contains many reservoirs, it will not be practicable to test each one for leakage, and a representative sample must be tested. Similarly, leakage from transmission mains is measured by choosing suitable representative sections. In a system which has only one or two major supply reservoirs serving the whole of a distribution network it may not be possible to measure reservoir leakage without interrupting supply, unless the reservoir has two chambers which can be tested independently.

2.9.1   Reservoirs

Leakage from reservoirs is measured by conducting a drop test. The aim is to measure the rate of fall of the water level over the duration of the test, with both the reservoir inlet and outlet valves closed. Operators must ensure that all inlet/outlet valves are drop tight. Ground around the reservoir should also be monitored for wet patches. The test should be done at night when demand is at a minimum. It may be possible to rearrange zone boundary valves to supply the zone with its reservoir under test from an adjacent zone.

The duration of the test depends on local supply criteria, but should be at least 4 hours and preferably 12 hours or longer. The reservoir should be full before the start of the test. The rate of drop can be measured in several ways:

a)   graduated scale
b)   depth gauge, which emits an audible signal on contact with the water level at
      each measurement
c)   pressure transducer and data logger; the pressure transducer is lowered to the
      bottom of the reservoir, and the data logger records the pressure drop in m head
      on the transducer by the reduction in water level
d)   capacitance rod: analogue signal depending on the depth of water submerged

The larger the surface area of the reservoir the smaller will be the rate of drop, if
any. This should be borne in mind when selecting the measurement instrument and
its sensitivity.

$$\text{Rate of leakage} = \frac{(d_1 - d_2) \times A}{T} \, m^3 / hour$$

where d_1 = initial depth (m)
       d_2 = final depth (m)
       A = surface area (m^2)
       T = test duration (hours)

It is recommended that reservoir levels are monitored for at least 48 hours
before planning the test, to determine the time when the reservoir is at its fullest
level, and the time when water demand on the reservoir increases rapidly. Knowing
these times will determine when the test should be started and finished without
disrupting supplies to consumers. If the reservoir does not fill sufficiently it may
be necessary to change the supply arrangements to gradually increase reservoir
storage over several days prior to the test. However, if the reservoir is by-passed
during the test, pressures in the network should be monitored – an increase in
pressure could result in many new leaks and bursts.

Reservoir overflows can be monitored by noting if a container placed at the
overflow weir is filled, or by placing a straw up the overflow pipe (which is displaced
by flow). A test of flow over time can then be carried out.

2.9.2   Transmission (trunk) mains

These are the major transmission mains between sources and distribution areas,
with relatively few connections from them to the secondary system.

Most of the current techniques for measuring flows in transmission mains are
based on changes in velocity between two points along the main (pairs of insertion
turbine or electromagnetic meters) or changes in total volume flow between two
points (by pass meter, transmission main DMA, ultrasonic meters). Both techniques
are prone to errors.

The techniques are summarised in Table 2.2. Those which require the main to
be taken out of service are identified.

Table 2.2 Summary of transmission main measurement technologies

	Main in service
Pairs of insertion meters	Yes
Ultasonic meters	Yes
Meter on by-pass	No
Transmission main DMA	Yes
Dilution gauging	Yes

Using temporary meters

This is a procedure for measuring flows and pressures in the distribution network where there are no existing meters or where the accuracy of existing meters requires checking. The method involves the use of an insertion probe meter, such as a pitot tube, an insertion turbine meter, or an insertion electromagnetic meter, to measure point velocity and derive flowrate. An acceptable alternative to such intrusive measurement techniques would be to use a clamp-on ultrasonic meter.

Insertion flowmeter installation

The installation technique is the same for both insertion turbine and electromagnetic meters. The meter is introduced into the live main through a 25 mm –50 mm gate valve or ball valve. The method used for drilling and tapping the main in order to fit a valve depends on the size of the main, and is illustrated in Figure 2.4

Figure 2.4   Drilling a main under pressure and installing a gate valve.

A split collar, which has been drilled and tapped to 50 mm thread is fitted to the main and a 50 mm nipple inserted in the hole. A 50 mm valve is attached to the nipple. An under-pressure drilling machine is attached to the valve, the valve opened, and a hole drilled in the main using a 37.5 mm inch drill bit. The drill is withdrawn into the pressure housing of the drilling machine, the valve closed and the drilling machine removed. The housing of the insertion meter is then screwed into the valve. When the valve is opened the probe and stem of the meter can be pushed under pressure into the centre of the main, as illustrated in Figure 2.5.

As an alternative to the split collar method, a stainless steel pipe repair clamp can be used, which incorporates a female threaded boss welded to the upper section of the clamp. A male threaded valve is attached to the boss, and the main drilled as previously described.

Flow and velocity measurements are recorded by means of a suitable data logger. The data logger should be programmed to measure flow at 5 or 10 minute intervals over the period of the flow survey. The data analysis program requires the internal diameter of the pipe in mm and the calibration factor of the meter in pulses/m, which is supplied by the manufacturer. Suitable data loggers are discussed in Chapter 6.

Pairs of insertion meters

To use this method, the following conditions must be met:

(i)   two insertion meters and two data loggers are available
(ii)   suitable tapping points can be made on the main through which to insert a flowmeter

Figure 2.5 Insertion meter and data logger in position.

(iii)   branch connections can be isolated or metered during the test period
(iv)   the flowrate along the main can be varied, e.g. by throttling a valve

Two insertion flowmeters are required, one inserted at each end of the section of main under test. The test period is two days. After the first day the meters are interchanged to eliminate inaccuracy from meter error. The leakage rate is calculated when the data loggers are read back at the end of the test.

UK water companies use insertion meters routinely as master meters for verification of production meters. The electromagnetic (EM) insertion meter is the most widely favoured. Most practitioners believe that the EM probe is less dependent on site conditions for accuracy than the insertion turbine meter. However one water company in the UK uses insertion turbine meters as well as insertion EM meters for bulk meter verification. It asserts that the insertion turbine meter can have a high level of accuracy when used in two planes, with a 20-point profile, and if care is taken in choosing the installation point.

Using pairs of insertion meters – i.e. one at each end of the main, should therefore still be a viable technology to monitor differences in velocity and thereby leakage. The method is not widely favoured because of the accuracy limitations, although the errors can be reduced by exchanging meters and averaging velocities. This is described in 'Report 26' [14].

Pairs of portable (clamp-on) ultrasonic meters

Portable (clamp-on) ultrasonic meters have been trialled by several UK companies, as an alternative to insertion meters, both for measuring production and for meter verification. UK water companies are increasingly using portable ultrasonic meters to measure flows out of reservoirs, but not in pairs to monitor leakage. One company had experimented with a similar technique – exposing the main at intervals along a transmission main and measuring flows at each point with a clamp-on meter. The UK petrochemical industry, however, regularly uses the technique for monitoring losses in their pipelines. Transducers are spaced at 600 m intervals – an accuracy of ±1% is claimed.

Managing Leakage Report J – Techniques, Technology and Training [1], describes a pilot exercise in France for permanently monitoring large diameter strategic pipes, e.g. motorway crossings, to enable fast follow-up action on bursts. Ultrasonic transducers were installed at each end of the main, with acoustic loggers at intervals in between. Monitoring was therefore at two levels: comparison of flow rate and acoustic monitoring. The former is suitable for detecting bursts, the latter for detecting small leaks as they occur.

There is no reason why such a technique should be less accurate than pairs of insertion meters, once the characteristics of the pipe have been accurately assessed and entered. However one UK company notes that results from a meter verification exercise at the same site, done by three different operators, showed an error range of 4%–20%.

Water mains also suffer from a build up of encrustation, reducing the accuracy of temporary insertion and ultrasonic measurement techniques.

Temporary full-bore meter

There is no reason why a full bore mechanical meter cannot be cut into a main for temporary flow measurement, except that, unlike insertion or clamp-on ultrasonic meters, the main must first be taken out of service.

BY-PASS METHOD

To use this method the following conditions must be met:

a)   the main can be taken out of supply
b)   there are valves on the main
c)   the valves can be closed and are drop-tight
d)   there are no branch connections from the main, or if there are, they can be turned off during the test

The procedure is as follows:

i)   Select a length of main to be tested. The length will depend on the distance between valves, but should be between 1 and 5 km.
ii)   Isolate the section of main under test by closing a valve at each end of the test section. Ensure that each valve is drop-tight.
iii)   Install a 25 mm positive displacement meter on a bypass around the upstream valve.
iiv)   Record meter readings before and after the test period. The test period should be as long as possible, within the constraints of system operation, and preferably at least one hour.

Any leakage from the section under test will be recorded by the meter. Leakage is calculated by the following formula:

$$\text{Leakage in litres/hour} = \frac{\text{flow through meter overtest period (litres)}}{\text{test period (hours)}}$$

Transmission main leakage can be expressed in litres/km/hour. The test can be repeated over several lengths of main.

Transmission (trunk) main district meter area

The question of whether to include transmission mains in the district meter area (DMA) system is addressed by companies in different ways. Some companies include transmission mains, particularly those with multiple branch connections, within DMAs, with flow measured at a DMA meter installed at each end of the transmission main at the DMA boundary. In this case transmission main leakage is difficult to separate from total DMA leakage, and the transmission main is subject to leak detection activities with the rest of the DMA, when monitoring shows and increase in night flow.

Transmission mains with few connections can be treated as a 'transmission main DMA', with a flowmeter at each end. Meters are usually full-bore mechanical or electromagnetic. Leakage monitoring is subject to the errors on each meter, and to changes in use by customers on branch connections. However, if branch connections can be shut off for regular monitoring exercises, meter accuracy is less significant than repeatability. As with other DMAs, regular night flow monitoring provides an indication of a burst or increasing leakage. The demand pattern on the transmission main inlet meter should be monitored and compared with the demand pattern of all the branch connections. If the difference is consistent over 24 hours, or it varies with pressure rather than demand, it is likely to be leakage. If it varies with total demand it is likely to be meter discrepancy.

Some companies exclude transmission mains from the leakage monitoring system altogether, metering at the branch connections only. Such mains are usually subjected to only occasional leak detection surveys.

Dilution gauging

Dilution gauging is a flow measurement technique which was developed by WRc, and which is still available via contractors. Although the technique has been used in the past, because of the analytical process required it is considered unwieldy and unsuitable for routine monitoring.

Pressure testing

This method applies particularly to new mains, with the main out of service. The ends of the main are capped and water is pumped in until fully charged. If the main leaks the pressure will fall – the rate of inflow needed to maintain pressure in the main is an indication of the leakage volume.

2.9.3   The distribution network

A flow measuring system in a water distribution network would ideally encompass measurement of total flows, to assist demand prediction and distribution management, and also zonal flows, which allows the engineer to understand and operate the system in smaller areas, and allows leakage management and control to take place. The system must therefore be hierarchical, i.e. at a number of levels, beginning at production measurement and ending at the customer's meter or consumption estimate. The system comprises:

- measurement of production at the source or treatment works
- measurement of flow into supply zones, with geographic or hydraulic boundaries
- flow monitoring into district meter areas of 500–3000 properties, with permanently closed boundary valves
- small leak location areas within each DMA, of around 500–1000 connections, where boundary valves remain open except during a leak location ('step test') exercise
- individual consumer meters, both domestic and commercial

Because of the size and complexity of the pipework in the distribution network, it is not always possible to measure leakage directly, except in small networks fed by a single metered feed. Therefore leakage in the distribution network is derived from measurement of night flows into the system or part of the system. Measurement is made by conducting a drop test (in a reservoir supply zone) or by selecting a representative area within the distribution system, and isolating the zone by closing valves around the boundary. A meter is installed on the main supplying the zone and the night flowrate is recorded. Consumption of large metered consumers is monitored over the same period and deducted from total night flow to give net night flow.

The drop test method is similar to the method for measuring reservoir leakage, except that the reservoir outlet valve is open. Zone valves should be checked for drop-tightness, and the reservoir level monitored during the night. The fall in level is an indication of genuine night consumption and leakage. It can be assumed that genuine night consumption is very small, around 1.7 litres/property/hour in the UK (from Managing Leakage – Report E [1]), during the time of minimum night flow (usually between 0200 and 0400).

In those systems where supply is intermittent, supply arrangements may be changed temporarily to ensure that the zone under test receives a continuous supply. The steps required for diversion of supply are explained in section 6.9.1. It is advisable to allow the system to equilibrate for 48 hours, for tanks to be filled, etc. before the night flow measurements are taken.

Flows into selected zones can be monitored using:

- existing zone meters
- meters installed specially for the test
- temporary insertion probe meters

2.10   REFERENCES

1   WSA/WCA Engineering and Operations Committee (1994) *Managing Leakage: UK Water Industry Managing Leakage* Reports A–J: Report A – *Summary Report*; Report B – *Reporting Comparative Leakage Performance*; Report C – *Setting Economic Leakage Targets*; Report D – *Estimating Unmeasured Water Delivered*; Report E – *Interpreting Measured Night Flows*; Report F – *Using Night Flow Data*; Report G – *Managing Water Pressure*; Report H – *Dealing With Customers' Leakage*; Report J – *Techniques, Technology and Training.* London: WRc/WSA/WCA.

2   Lai, Cheng Cheong (1991) *Unaccounted for Water and the Economics of Leak Detection*, IWSA International Congress, Copenhagen. IWSA International 'Special Subject' Report 1. London: IWSA.

3   Hirner, W and Lambert, A (2000) *Losses from Water Supply Systems: Standard Terminology and Recommended Performance Measures.* London: IWA.

4   OFWAT (2001) *Leakage and the Efficient Use of Water*, 2000–2001 report. Birmingham: OFWAT.

5   Water Services Association of Australia (2001) *Benchmarking of Water Losses in Australia, User Manual.* Melbourne: WSAA.

6   2nd World Water Congress – Water Distribution and Water Services Management. (2001) Proceedings. London: IWA.

7   Malta Water Services Corporation (2001) *Annual Report 2001.* Valetta: Malta Water Services Corporation.

8   New Zealand Water & Wastes Association (2002) *Benchmarking of Water Losses in New Zealand (Incorporating the User Manual for the BenchlossNZ Software: Version 1A). User Manual.* Wellington: New Zealand Water & Wastes Association.

9   McKenzie RS, Lambert AO, Kock JE and Mtwshweni W (2002) *User Guide for the Benchleak Model, developed through South African Water Research Commission.* Report TT 159/01. Pretoria: WRC.

10   Global Water Resources Ltd (2002) Fastcalc 2002: customised software for rapid calculation of IWA Water Balance and Performance Indicators with 95% confidence limits. Global Water Resources Ltd.

11   Lambert, AO (1994) 'Accounting for losses: the bursts and background concept (BABE)', *IWEM Journal*, April, 8(2), 205–14.

12   Hirner, W and Sattler, R (2001) 'Failure and rehabilitation rates of distribution systems', paper presented at IWA Conference on System Approach to Leakage Control and Water Distribution Systems Management, Brno, 16–18 May 2001, Brno, Czech Republic. Brno: Brno University of Technology.

13 Lambert, AO (2001) 'International Report: Water losses management and techniques, (incorporating Portuguese National Report)', 2nd World Water Congress: Water Distribution and Water Services Management, Berlin, Germany, 15–19 October. *Water Science and Technology: Water Supply*, 2(4) 1–20.
14 Technical Working Group on Waste of Water (1985 [1980]) *Leakage Control Policy and Practice*, Standing Technical Committee Report no. 26. Original publication London: Doe/NWC. Reprinted London WAA/WRc.

3

International comparisons

3.1 PERFORMANCE INDICATORS AND TARGET SETTING

There are many possible performance indicators (PIs), but some are better than others, and many are unsuitable for some purposes. This was acknowledged in a report commissioned by the International Water Supply Association in 1991 [1], where comparisons of water losses in terms of percentages, per kilometre of mains and per service connection were presented and discussed, without any firm conclusions or recommendations. However, the next IWA international report on water losses management and techniques [2] produced some ten years later, contains much improved guidance on selection of PIs for particular purposes.

To take a further example, when monitoring night flows to identify the presence of unreported leaks, the measurement is usually taken in litres/second or m³/hour. For the purpose of selecting sectors to choose for leak location activities, the night flow rates would traditionally be converted to one of the following measures and compared:

- percentage of average daily flow (typical of some USA suppliers, and French international practice)

- m³/km of mains/hour (typical of current German and Japanese practice)
- litres/property/hour (typical of recent UK practice)
- litres/service connection/hour

The fact that these simple measures are based on use of some convenient easily identifiable parameter, however, does not necessarily mean that that parameter forms a suitable basis for making decisions on when to intervene to look for unreported leaks.

Following the introduction of component analysis techniques in the early 1990s, it is now considered preferable when using night flows for leak location to:

- identify a 'lowest achieved' night flow in litres/sec or m³/hour, after all detectable leaks have been quickly repaired or shut off
- compare the 'lowest achieved' to the lowest technical minimum flow rate which would be expected, given the likely background leakage (based on length of mains, number of service connections, average night pressure), and the customer night use (based on resident population, toilet cistern size, number and type of commercial properties etc)
- if the 'lowest achieved' night flow is close to the 'technical minimum', monitor the night flow and identify the excess in litres/second or m³/hour, and use that as the basis for deciding when to intervene to locate the excess.

3.2   THE IWA PERFORMANCE MEASURES STRUCTURE

Performance Indicators for Water Supply Services: IWA Manual of Best Practice [3] contains 133 different performance indicators for different functions – water resources, personnel, physical, operational, quality of service, and financial. Each function can also have up to four levels of indicator, according to their importance as management tools.

Performance indicators range from level 1 (basic), which provide a general management overview of efficiency and effectiveness, to level 3 (detailed) indicators, which deal with specific elements of operational management.

3.3   NON-REVENUE WATER: FINANCIAL PERFORMANCE INDICATORS

The IWA 'best practice' level 1 performance indicator for non-revenue water (NRW) is:

Volume of non-revenue water as a percentage of system input volume [3, Fig. 36].

International data for the percentage of NRW by volume [1, 4] typically show the percentage of NRW varying from less than 5% to over 50%. There are many reasons for this wide range of variation, other than simply management efficiency and infrastructure condition:

- economic NRW management policies depend upon the cost and availability of water
- high consumption decreases % NRW, and low consumption increases % NRW
- intermittent supply reduces the length of time the system is pressurised and leaking, but is not good practice as it reduces infrastructure life
- apparent losses are influenced by type of meters, and whether customers are supplied by direct pressure or via roof tanks
- average operating pressures vary from less than 20 m to over 100 m, and average real losses vary approximately linearly with pressure for large systems with mixed pipe materials
- some systems include transmission mains and service reservoir real losses, others do not
- real losses may include leakage on customers private pipes, depending on ownership and maintenance responsibility for different sections of the service connection, and customer meter location

NRW has, for many years, been quoted only or principally in percentage terms. Accordingly many non-specialists, including politicians and the media, incorrectly believe that this is the most meaningful measure of performance for NRW and all its components. So targets are often set, or suggested, at national level in percentage terms. Whilst this is undoubtedly better than setting no targets at all, it discriminates against utilities with low consumption (low system loading), higher than average operating pressures (due to topography), and NRW calculations which include leakage on customers' private pipes.

Technical groups in Germany (DVGW) and the UK have, for many years, drawn attention to the undue influence of consumption, and changes in consumption, when water losses are expressed as a percentage of system input volume. The UK economic regulator, the Office of Water Services (OFWAT), and the South African Bureau of Standards have more recently decided against continued use of percentages for making performance comparisons of real losses. The IWA 'best practice' report [3] specifically states that percentages by volume are unsuitable for assessing performance in operational management of real losses.

The undue influence of consumption, and changes in consumption, is demonstrated in Figure 3.1. The x-axis shows the consumption per service connection per day, ranging from as low as 250 litres/connection/day (Malta) to over 8000

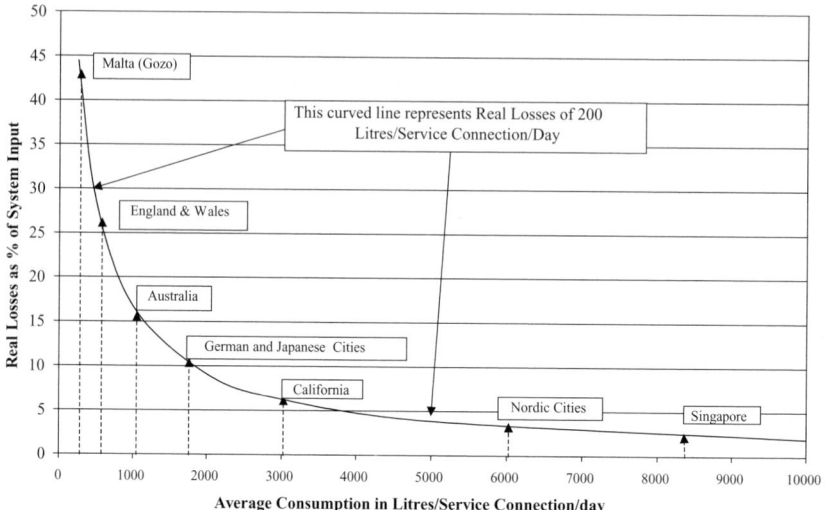

Figure 3.1 The influence of consumption on real losses expressed as
percentage of system input volume.

litres/connection/day (Singapore). The curved line represents real losses of 200
litres/service connection/day – the average of the recent international data set
reference [5].

Depending upon the consumption per service connection, the same volume of
real losses could, in percentage terms, be anything from 44% to 2.4%. Thus, countries
with relatively low consumption, such as Malta, England and Wales, and many
developing countries, can appear to have high losses when expressed in percentage
terms; in contrast, percentage losses for urban areas in developed countries with
high consumption can be equally misleading.

When consumption decreases, seasonally or annually, or due to demand
management measures, the percentage of real losses increases even if the volume
of real losses remains unchanged. When consumption increases, the opposite effect
occurs. A recent Polish National Report [6] demonstrates the same conclusion,
namely that percentage real losses decrease rapidly as system loading (in m³/km
mains/hour) increases.

There are also problems of interpreting percentage real losses in intermittent
supply situations, and auditing future targets for real losses stated in percentage
terms.

Despite these deficiencies, some practitioners still consider that percentages
can be used for comparing a particular utility's year-on-year performance. However,
it is by far preferable to make such simple year-on-year comparisons using a scaling

factor related to the distribution infrastructure, such as litres/service connection/ day.

The IWA Task Forces have also recommended a level three financial performance indicator for non-revenue water. For the calculation of this PI [3, Fig. 37] the volume of each of the main components of NRW is assigned a valuation in local currency/ m^3, appropriate to local circumstances, and the value of the NRW component is expressed as a percentage of the annual cost of running the system.

3.4 APPARENT LOSSES

The IWA recommended PI for apparent losses [3] is m^3/service connection/year. However, in systems where all customers are metered, and the component of illegal use/theft is small, it may be preferable to express apparent losses as a percentage of authorised metered consumption, as most of the apparent losses will be due to customer metering errors.

3.5 REAL LOSSES

3.5.1 The best basic performance indicator for operational management of real losses

The IWA recommendation for the 'best practice' level 1 performance indicator [3, Op 24] for operational management of real losses recognises that:

- percentage of input volume is too strongly influenced by consumption, and changes in consumption, making it unsuitable for this purpose
- 'per billed account' or 'per property' should not be used, as some service connections supply multiple billed properties, yet there is only one service connection with potential to leak
- the choice of 'per service connection' or 'per km of mains' as a scaling factor depends upon the density of connections for the system under consideration

Figure 3.2 shows this selection process for IWA PI Op 24 in the form of a decision diagram. As most distribution systems have density of connections greater than 20 per km of mains, 'per service connection' should logically become the predominant basic operational PI for real losses in future.

In the case of systems subject to intermittent supply, Op 24 is expressed as 'litres/service connection/day when the system is pressurised'. The annual volume of real losses is divided by the equivalent number of days that the system is pressurised, rather than by 365 days. Figure 3.3 shows the values of Op 24 for the Water Losses Task Force international data set [5].

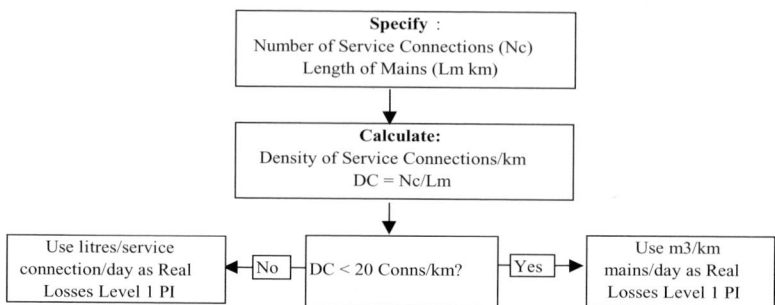

Figure 3.2 Process to determine level 1 PI for operational management of real losses.

Comparatively few countries currently publish real losses in litres/service connection/day, although some (including the UK) traditionally use litres/property/day. However, since the IWA 'best practice' report recommendations, an increasing number of utilities, countries, consultants and funding agencies are showing interest in using litres/service connection/day as the preferred basic PI for operational management of real losses in future [7].

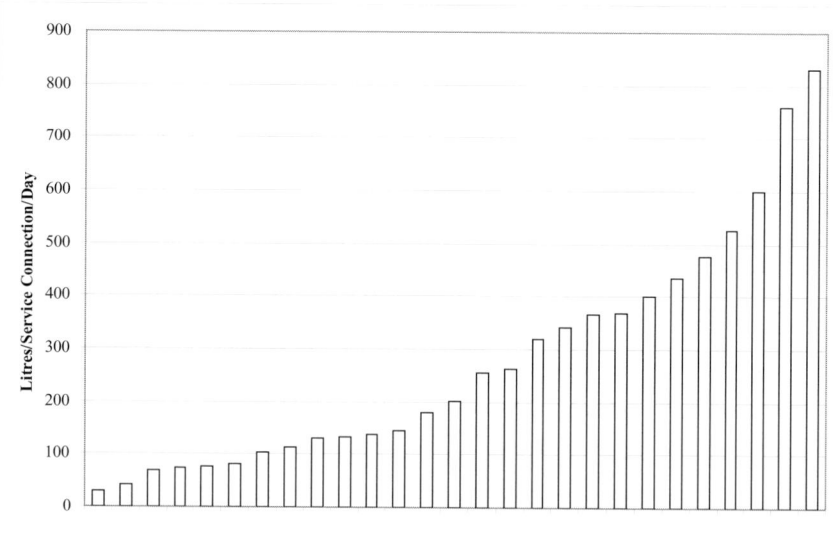

Figure 3.3 Real losses in litres/service connection day for international data set systems.

The level 1 PIs of 'per service connection' and 'per km of mains' provide basic comparisons of performance, but operational management of real losses is strongly influenced by differences in density of connections, customer meter location and average pressure. The IWA Task Force on Water Losses used component analysis techniques to investigate the influence of these three factors on unavoidable (technical minimum) annual real losses (UARL).

3.5.2   Unavoidable annual real losses

The Water Losses Task Force developed a system-specific equation [2] for the lowest technically achievable real losses. Using appropriate parameter values for burst frequencies, maximum durations and pressure-related typical flow rates for well-managed systems with infrastructure maintained in good condition, the following components of UARL were obtained.

On mains:		18 litres/km mains/day/metre of pressure
Plus	On service connections (up to property boundary)	0.8 litres/service connection/day/ metre of pressure
Plus	On service connections (property boundary to customer meter)	25 litres/km/day/metre of pressure

UARL is a useful concept as it can be used to predict, with reasonable reliability, the lowest technically achievable annual real losses for any combination of mains length, number of connections, customer meter location and average operating pressure – assuming that the system is in good condition with high standards for management of real losses.

Figure 3.4 shows how the UARL varies with the density of connections (per km of mains), for systems with customer meters located close to the property boundary. If UARL is expressed in m^3/km of mains/day/metre of pressure (right hand axis), the UARL value rises rapidly as density of connections increases; the value at 120 connection/km being over three times the value at 20 connections per km. This means that when real losses are expressed 'per km of mains', it is only possible to compare performance (or set targets) for systems within specified narrow bands of density of connections. The IWA German National Report [8] clearly demonstrates this constraint.

Alternatively, if UARL is expressed in litres/service connection/day/metre of pressure (left hand axis), the UARL value is almost constant for a wide range of density of connections from 60 per km upwards, at 1 litres/service connection/day/ metre of pressure ($\pm10\%$). As density of connections decreases from 60 to 20 per

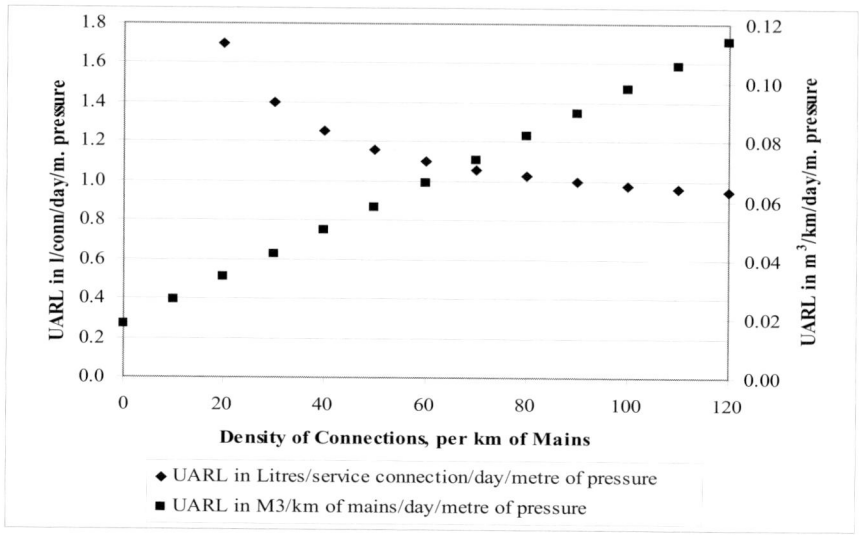

Figure 3.4 UARL and density of connections.

km of mains, the UARL in these units rises by around 50%, and this measure is not recommended for connection densities less than 20 per km of mains.

In Figure 3.5, suppose that the area of the large rectangle represents the current annual real losses, in m³/year, for any specific system. As the system ages, there is a tendency for a natural rate of rise of real losses through new leaks and bursts, some of which will not be reported to the utility. This tendency is controlled and managed by some combination of the four primary components of real losses management, namely:

• pipeline and assets management
• pressure management (which may mean increases or decreases of pressure)
• speed and quality of repairs
• active leakage control, to locate unreported leaks

The number of new leaks arising each year is influenced primarily by long-term pipeline management. Pressure management can influence the frequency of new leaks, and the flow rates of all leaks and bursts. The average duration of the leaks is limited by the speed and quality of repairs, and the active leakage control strategy controls how long unreported leaks run for before they are located. The extent to which each of these four activities is carried out will determine whether the volume of annual real losses increases, decreases or remains constant.

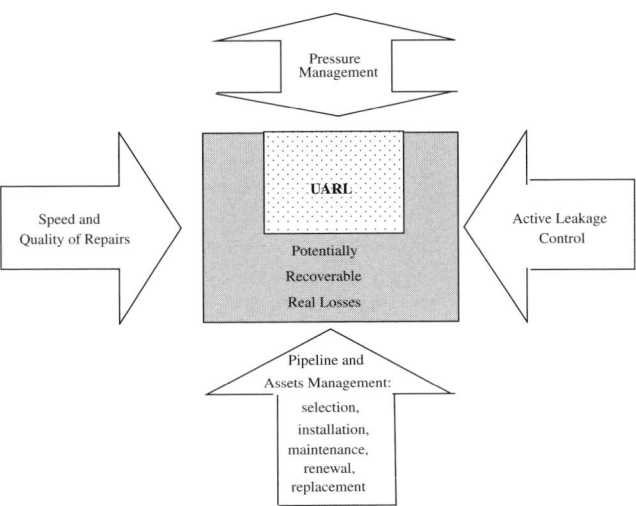

Figure 3.5 The four basic methods of managing real losses.

Real losses cannot be eliminated totally. The lowest technically achievable annual volume of real losses for well-maintained and well-managed systems is the UARL, represented by the smaller rectangle in Figure 3.5. System-specific values of UARL can be calculated using the component-based methodology developed by the Water Losses Task Force and described previously.

The difference between the UARL (small rectangle) and the current annual real losses (CARL) is the potentially recoverable real losses. The ratio of the CARL to the UARL is the infrastructure leakage index (ILI).

The ILI measures how effectively the infrastructure activities in Figure 3.5 – repairs, active leakage control and pipeline/assets management – are being managed at current operating pressure.

For each of the four activities, there is some economic level of investment and activity, which needs to be calculated or assessed, depending upon the marginal value, in local currency/m^3, placed on the real losses. Depending upon local circumstances and practice, the marginal value placed on real losses may be low – perhaps power and chemicals cost only – or high, and this profoundly influences the economic management policies for controlling real losses.

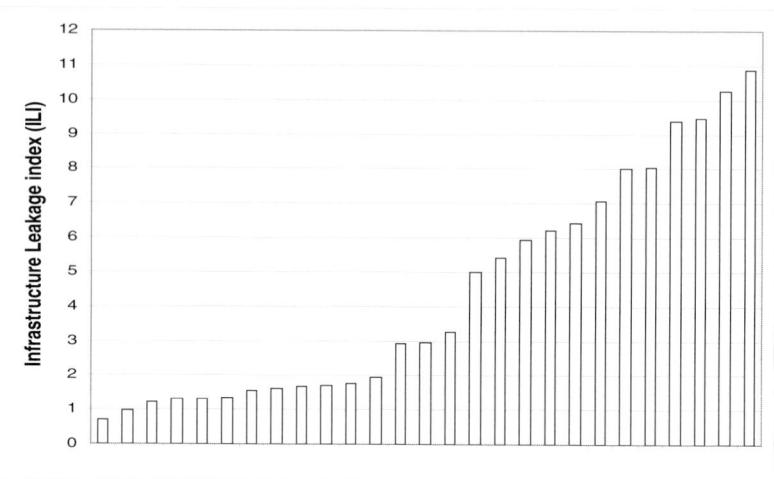

Figure 3.6 Infrastructure leakage index for the international data set systems.

3.5.3   A detailed performance indicator for operational management of real losses

The detailed (level 3) PI recommended by the IWA for operational management of real losses is the infrastructure leakage index, which is the ratio of the current annual real losses to the unavoidable annual real losses shown in Figure 3.5. The basis of calculation of the UARL makes due allowance for length of mains, number of service connections, location of customer meters and average operating pressure. The ILI measures how effectively the three infrastructure activities in Figure 3.5 – speed and quality of repairs, active leakage control and pipe materials – are being managed, at the current operating pressure.

Figure 3.6 shows calculated values of ILI for the international data set. Values close to 1.0 represent near-perfect technical management of real losses from infrastructure, at actual operating pressures.

Since the ILI was introduced as a performance indicator in 1999 [3] it has generated significant interest as it can be used for performance comparisons for systems exceeding 5000 service connections, 25m pressure and 20 conns/km at utility, national and international level.

National organisations in South Africa [9], Australia [10], New Zealand [11] and Malta (Case Study 4) have commissioned software and training manuals for introducing and applying the ILI. The Leak Detection and Water Accountability Committee of the American Water Works Association (AWWA) has recommended that the methodology be introduced throughout AWWA and North America, and several papers have been written describing initial performance comparisons using the ILI approach. Further papers on this topic were presented at an IWA Conference in Cyprus – Leakage Management, a Practical Approach – in November 2002, and are available from the IWA website.

3.6 PERFORMANCE INDICATORS IN ENGLAND AND WALES

The ILI methodology has been used by the UK National Audit Office [12] to compare leakage management performance in England and Wales with international data, using the data set in Figure 3.6. The ILI methodology has also been used by several water companies to compare real losses for the many diverse supply systems they operate.

Since 1995, OFWAT has published annually the independently audited figures for distribution losses for water companies in England and Wales. Recognising the influence of changes in consumption, as a matter of policy OFWAT has made a deliberate decision not to show figures for losses in percentage terms.

The annual OFWAT reports show total losses (distribution losses plus supply pipe losses) in volume terms (Ml/day), in litres/property/day, and in m^3/km of mains/day. By 2000/01, most companies [13] had achieved (or almost achieved) their mandatory leakage targets.

At company level, density of connections in England and Wales ranges from around 50 to 100 per km of mains. Environment Agency estimates of average pressure range from 25 m to 56 m (median 42 m). Private supply pipes average around 15 metres length. Using these figures, informal calculations of ILIs for the companies reaching their economic leakage targets appear to be in the range 1.4 to 3.0, with a median value of around 2.0.

It would be expected from Figure 3.4 that, for density of connections ranging from 45 to 100, 'litres/service connection/day' will be a more robust simple performance for real losses than 'm^3/km of mains /day' (which is more strongly influenced by differences in density of connections).

This brief example shows how the IWA studies could assist individual countries in improving their choice of performance indicators. In the case of England and Wales, and other countries such as Malta having relatively high connection densities at company level, simple comparisons of real losses in m^3/km mains/day are more likely to confuse than to inform.

Further enlightenment could be achieved by calculating ILI values, or as an intermediate step, as shown in the following example, taking account of average pressure by comparing real losses in 'litres/service connection/day/metre of pressure'.

A near linear relationship between leakage rate and pressure for aggregated sector data has recently been confirmed in a major UKWIR research project. Table 3.1, loosely based on actual UK situations, shows how the perspective on leakage management performance can change markedly when pressure is taken into account. System X is located in a flat area with low pressures; System Y in a hilly area with quite high pressure, even after extensive pressure management. Without allowing for average pressure, real losses in System Y (in litres/service connection/day) appears to be around 50% higher than real losses in System X. Allowing for average operating pressures, it is seen that real losses in System Y are in fact 33% lower (not 50% higher) than in System X.

Table 3.1 How inclusion of average pressure changes simple perception of leakage management performance

System	Real losses (litres/conn/day)	Average pressure (metres)	Real losses (litres/conn/day/metre of pressure)
X	80	25	3.20
Y	120	56	2.14

3.7   REFERENCES

1   Lai, Cheng Cheong (1991) *Unaccounted for Water and the Economics of Leak Detection*, IWSA International Congress, Copenhagen. IWSA International 'Special Subject' Report 1. London: IWA.

2   Lambert, A (2002) 'International report on water losses management and techniques', *Proceedings of IWA Congress, October 2001, Berlin. Water Science and Technology: Water Supply* 2(4), London: IWA.

3   Alegre, H, Hirner, W, Baptista, JM and Parena, R. (2000) *Performance Indicators for Water Supply Services: IWA Manual of Best Practice*. London: IWA.

4   Lallana C, Krinner W, Estrela T, Nixon S, Leonard J, Berland JM, Lack TJ, Thyssen N. (2001) *Sustainable water use in Europe: Part 2: Demand Management*. Environmental Issue Report No 19: European Environment Agency.

5   Lambert AO, Brown TG, Takizawa M and Weimer D (1999) 'A review of performance indicators for real losses from water supply systems', *AQUA*, 48(6), 227–237.

6   Lambert, AO (2001) 'International Report: Water losses management and techniques, (incorporating Polish National Report)', 2nd World Water Congress: Water Distribution and Water Services Management, Berlin, Germany, 15–19 October. *Water Science and Technology: Water Supply*, 2(4) 1–20.

7   Lambert, AO and McKenzie, RD (2002) 'Practical experience in using the infrastructure leakage index', paper presented at the IWA Conference 'Leakage Management – a Practical Approach', Cyprus, November.

8   Lambert, AO (2001) 'International Report: Water losses management and techniques,

(incorporating German National Report)', 2nd World Water Congress: Water Distribution and Water Services Management, Berlin, Germany, 15–19 October. *Water Science and Technology: Water Supply*, 2(4) 1–20.

9   McKenzie RS, Lambert AO, Kock JE and Mtwshweni W (2002) *User Guide for the Benchleak Model, developed through South African Water Research Commission.* Report TT 159/01. Pretoria: WRC.

10   Water Services Association of Australia (2001) *Benchmarking of Water Losses in Australia. User Manual.* Melbourne: WSAA.

11   New Zealand Water & Wastes Association (2002) *Benchmarking of Water Losses in New Zealand (Incorporating the User Manual for the BenchlossNZ Software: Version 1A). User Manual.* Wellington: New Zealand Water & Wastes Association.

12   UK National Audit Office (2000) Leakage and Water Efficiency: 2000 report, London: UK National Audit Office.

13   OFWAT (2001) *Leakage and the Efficient Use of Water*, 2000–2001 report, Birmingham: OFWAT.

4

Developing a strategy

4.1   INTRODUCTION

Much that has been written about leakage economics, and from that the development
of a strategy, has been aimed at either the leakage specialist or the economic regulator,
or it has been designed to meet an academic audience interested (though not directly
involved) in the theory of leakage economics.

This chapter aims to make the practitioner aware of the key messages from the
theory and the issues, which have to be taken into account, when developing
appropriate policies and practices. The tasks and solutions involved in developing
a strategy are illustrated in Figure 4.1, and the detail of policy implementation is
described in subsequent chapters.

The most important aspect of any leakage strategy is the leakage target. What
level of leakage should the water supplier aim for, and what level should be
maintained in the longer term?

Leakage is synonymous with waste. Therefore, a company which wastes the
product it is meant to supply, must come under scrutiny to investigate what else it
wastes, for example customer income. This in turn can lead to prices for the product
in excess of what is reasonable.

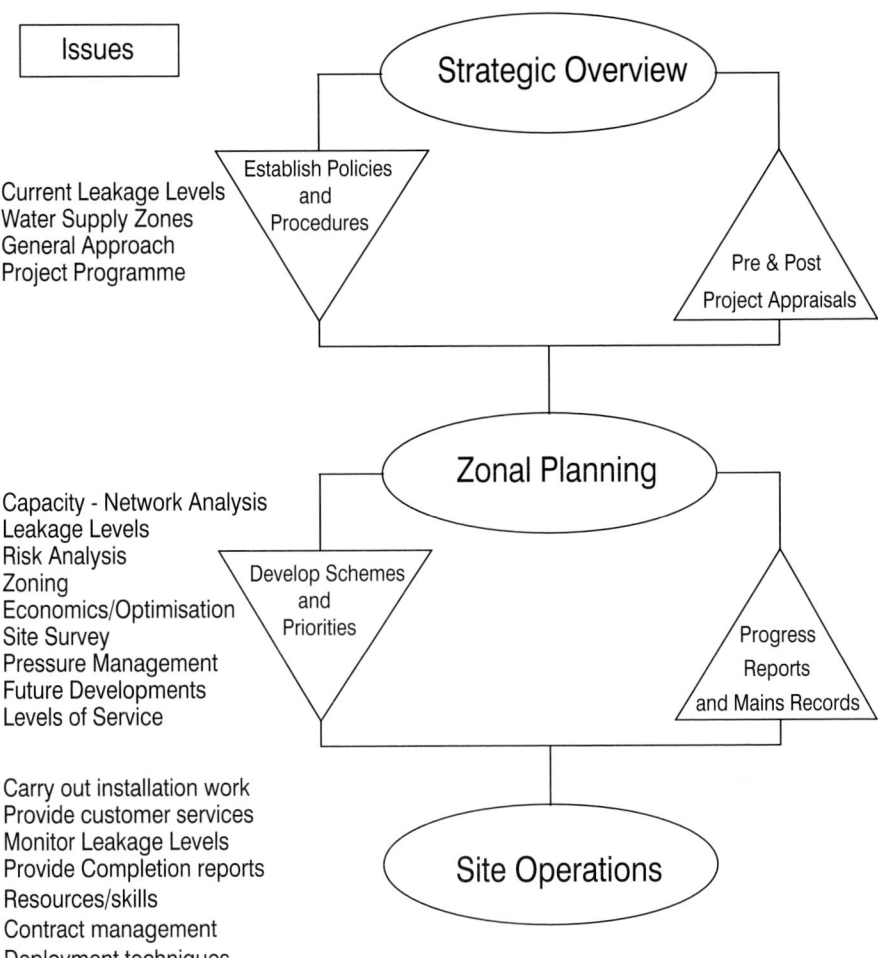

Issues

Strategic Overview

Current Leakage Levels
Water Supply Zones
General Approach
Project Programme

Establish Policies
and
Procedures

Pre & Post
Project Appraisals

Zonal Planning

Capacity - Network Analysis
Leakage Levels
Risk Analysis
Zoning
Economics/Optimisation
Site Survey
Pressure Management
Future Developments
Levels of Service

Develop Schemes
and
Priorities

Progress
Reports
and Mains Records

Carry out installation work
Provide customer services
Monitor Leakage Levels
Provide Completion reports
Resources/skills
Contract management
Deployment techniques

Site Operations

Figure 4.1 A structured approach to leakage management.

4.2   ECONOMICS OF WATER LOSS

4.2.1   Diminishing returns

In an ideal world, every water supplier would like to eliminate leakage from water distribution systems. It serves no purpose, is wasteful, and has a deleterious effect on several aspects of the operation of a well-run water supply network. Leakage will add to the cost of producing and distributing water. It will add to the capacity requirement for storage systems, treatment works, and mains sizing.

However, just as we would like to eliminate world hunger, poverty and disease, we must be pragmatic and accept that it is an impossible task. For the vast majority of water distribution systems, leakage is something which cannot be eliminated completely. *There will always be a level of leakage which has to be tolerated, and which has to be managed.*

When considering alternative ways of bridging the gap between the future need for water into an area, and the current availability of water, there are usually two principal methods:

- supply augmentation – this may mean adding reservoir or pumping capacity, increasing treatment capacity, bringing in water from an adjacent area
- reducing the future need for water by leakage reduction and demand management

Each of these methods will increase the forecast 'headroom' i.e. the difference between the available supply and the projected demand, and therefore reduce the risk that supplies will become insufficient to meet customer demand. This is shown in Figure 4.2. However, when comparing the economics of the two methods there is a general principle which should be appreciated.

When building a new treatment works, there are economies of scale associated with the size of the plant. The larger the plant, the higher the cost; but as the size increases the unit cost (say in terms of the cost per Ml/d of capacity) tends to reduce due to efficiencies. A 100Ml/d plant will cost significantly less than double the cost of a 50Ml/d plant. There are certain costs e.g. those associated with design, contract administration, and land purchase which are less sensitive to capacity, than other costs.

This is not universally true however. There will be cases when the initial increases in supply can be achieved by extending an existing treatment works, whereas larger increases can only be met from a new source. In extreme situations, water may have to be derived from desalination processes at a considerable unit cost.

Reducing leakage costs money. However, unlike supply augmentation, there is less scope for economies of scale. In fact the reverse tends to be true. Everything to do with leakage reduction follows a law of diminishing returns; the more effort that is put in, the less will be the impact in terms of water saved.

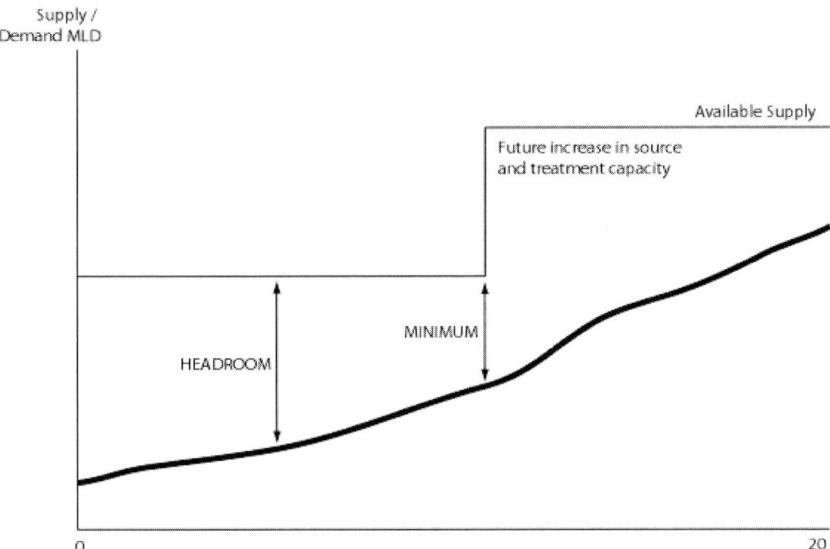

Figure 4.2 Water demand and supply headroom.

Figure 4.3 shows the prime techniques for leakage management, which have been referred to as the four 'pillars'. Each of these follows a similar law of diminishing returns.

Active leakage control

When first undertaking leakage detection and repair work, leaks will be relatively easy to find. A backlog may have built up due to under investment in previous years resulting in fewer leaks being found and fixed than occur in any given year. However, once the more obvious mains and service bursts have been found, then a higher level of effort has to be put in to reduce leakage by a similar volume.

Pressure management

The most cost effective schemes are those which cover a large area, and which make a significant impact on average pressures. An example would be the installation of a pressure reducing valve (PRV) on the branch from a trunk main to cover a whole town. Once such schemes have been completed, the next stage may be to install PRVs in conjunction with district metering. In the extreme, some water suppliers have installed PRVs on supplies to districts of less than 200 properties, or even on individual properties. The cost of the scheme reduces much less than the

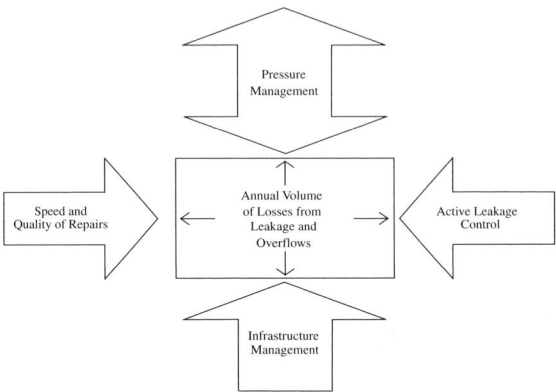

Figure 4.3 The four pillars of leakage management.

benefit obtained due to the reduction in the area covered, and so the schemes become less cost effective. The principle of designing pressure management areas (PMAs) is detailed in Chapter 7.

District metering

When installing zonal and district metering there is a tendency to favour areas which can be metered without the need for additional facilities to be installed in the network. Areas will be created using natural breaks in the network, along the lines of main roads, rivers and canals, and over undeveloped land.

The aim is to provide single feed district meter areas (DMAs) which are supplied through only one meter installation. This will tend to minimise the number of valves which have to be shut in order to create a discrete area. The number of areas which can be created in this way, will depend on the layout of the distribution network. The cost and the benefit will be similar.

As the number of properties contained within DMAs increases, there will come a point at which adding further properties incurs a higher unit cost. This may be due to:

• the need to install additional valves
• the need to provide two or more meters into the area
• the need to lay new lengths of link main

In the experience of the author (Trow), it is usually possible to create DMAs of between 50 and 70% of properties using the existing network, and so there will be a similar unit cost (per property) up to that level of coverage. The next step is to consider those DMAs which require new facilities such as valves and link mains. This can add a further 10 to 30% of properties to DMAs economically. However, there are usually some properties (5 to 15%) which cannot be added to DMAs without incurring costs for which there is no clear payback in terms of water saved. These are often in city centre areas where the demand and the types of property (e.g. high rise buildings, and fire sprinkler supplies) may require a number of mains in the same street, running at different pressures, which are interconnected for emergency use. The principle of zoning and dividing a zone into DMAs is detailed in Chapters 5 and 6.

Mains and service renewal

Replacing an old water main with a new installation, which is pressure tested before it is put into service, will undoubtedly reduce leakage on the main. Most leakage occurs on service connections and, unless the service connections are also renewed, the benefit may not be a great as first estimated.

In some cases, it may be difficult to transfer all the service connections from the old main to the new main, and so the old main is left in the ground. This will only add to the leakage problem and, even worse, the old main may be erased from the record plans, so no leakage detection will be carried out on it.

If water mains are being replaced for reasons other than leakage control, for example water quality problems, or due to customer service issues, then the benefit to the leakage engineer should also be taken into account. When mains replacement is being used as a primary measure for leakage reduction, targeting studies should be carried out to determine which areas, and which mains within those areas, have the highest burst frequency (number per kilometre per year) and which have the highest levels of background leakage.

If targeting is carried out effectively, it is inevitable that the initial schemes will be more cost effective than the later ones. So, mains replacement will also follow a law of diminishing returns.

Speed and quality of repairs

Reducing the time it takes to repair a leak will reduce the volume of leakage. However, once the repair time is reduced below a certain threshold, the unit repair cost will tend to rise due to standby, call out and overtime payments to staff, or supplementary payments to contractors to make additional repair teams available.

Figure 4.4 shows how each leakage management option can be implemented at a relatively low unit cost per Ml/d, but the unit cost increases the more each measure is used. The diagram shows a typical relative order for using each measure, but this order may vary depending on local circumstances. Although each option is illustrated with a set start and end level leakage, in practice there will be some degree of overlap, where a mixture of measures is used at any level of leakage to make further reductions.

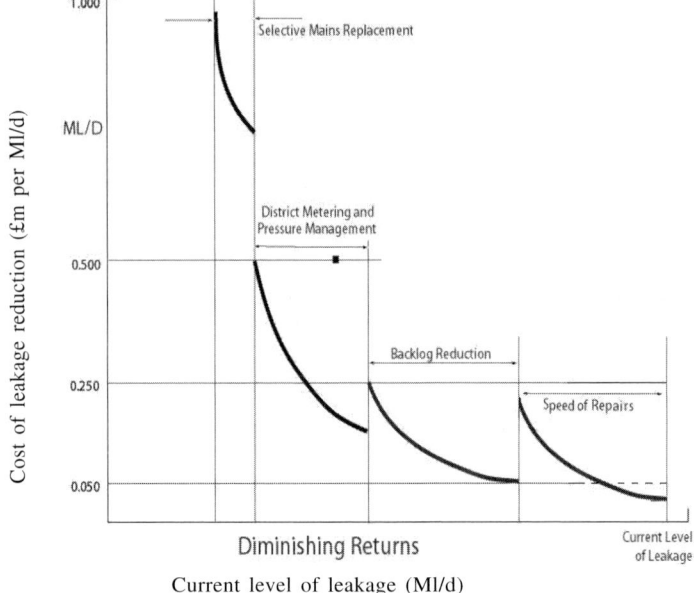

Figure 4.4 Diminishing returns from leakage management measures.

4.3   ECONOMIC LEVEL OF LEAKAGE

For any water distribution system there is a level of leakage below which is it not cost effective to make further investment, or use additional resources, to drive leakage down further. In other words, the value of the water saved is less than the cost of making the further reduction. This point is known as the economic level of leakage (ELL).

Much has been written about the development of the ELL theory and the methods of assessing the value. Some of the key points are highlighted here:

- There is no single ELL. The ELL will vary over time depending on factors such as seasonal changes to burst frequency, often resulting from weather conditions, mains condition improvements.
- Investments in pressure control, district metering and telemetry to reduce leakage will also change the ELL based on the degree of active leakage control (ALC) effort.
- The value of water will change over time. It will have higher value in times of shortage e.g. droughts, and a lower value in times of plenty. The value will increase as headroom is reduced, and it will fall when new sources and treatment

works come on line. The operating costs may change in future due to different water quality standards, or changes to regulations making current practices obsolete.

- New leak detection techniques will change the efficiency of detection operations resulting in a change to the true ELL. The ELL will be different depending on the method used for leak detection – e.g. a policy of regular inspection or one based on continual night flow monitoring in district meter areas.
- The estimation of ELL must use data, information and policy rules specific to that area and that water supply organisation. However, until significant work has been conducted to reduce leakage and so collect the necessary data on costs and effects, it is not possible to make an accurate assessment of ELL. Therefore, the calculation of ELL will follow a staged approach and could take several years to determine accurately.

Leakage targets based on ELL must therefore be specific and dynamic.

4.3.1   A brief history of leakage economics

The theory of an economic level of leakage is not new. Joseph Parry in his 1881 book *Water – Its Composition Collection and Distribution'* [1] discusses the costs and benefits from waste prevention by inspections, use of good quality fittings and materials, and early results form a system of waste metering in Liverpool. This was a time when most parts of London received 'an imperfect intermittent supply', and the priority was to achieve a constant supply. In 1957, Gledhill (then engineer and manager of Sutton District Water Company) investigated the economics of regular sounding and produced some of the theoretical relationships still used today [2].

In 1980, 'Report 26' [3] compared the cost benefit of various forms of leakage control from passive control, regular sounding, and waste metering. Report 26 stated: 'It is clearly uneconomic to ensure that pipelines and reservoirs will never leak. It is also clear that there is an economic limit to the loss of water which should be tolerated through leakage'. In 1988, DG Shore [4] proposed a method of target setting based on optimum cost calculations.

Managing Leakage – Report C [5], published in 1994, defined the ELL as 'that level of leakage where the marginal cost of active leakage control equals the marginal cost of the leaking water'. In other words, when the cost of reducing leakage by one cubic metre equals the value of that same or equivalent cubic metre of water.

In all these cases, the theory is similar. Figure 4.5 shows the generalised relationship between expenditure on leakage management operations, and the unit production costs of water as a function of the level of losses. The key to a successful strategy is to collect sufficient factual data to allow this relationship to be understood for each supply zone.

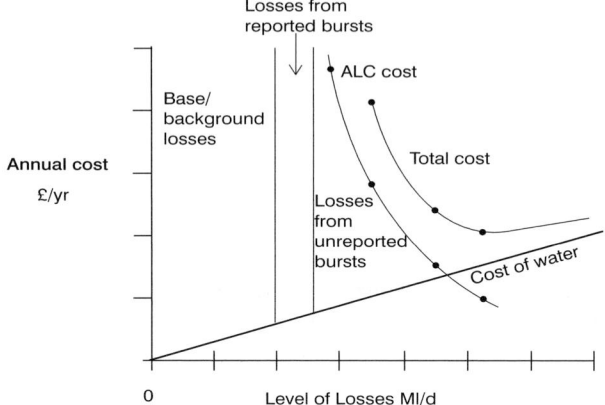

Figure 4.5 General relationship between operating costs and the level of losses.

4.3.2   Costs and value of water

In order to understand the estimation of ELL, it is necessary to appreciate how water is valued. This will vary from one region to another, and also within areas of the same region.

Short-term ELL

In the short term there are a number of key parameters, which govern the actual level of leakage. These are effectively fixed, and include:

- the average pressure in the system
- the condition of the mains and service pipes
- the facilities available for collecting data (i.e. district metering and telemetry)

So, at any particular time, the only parameter which can be changed quickly, to have an impact on the level of leakage, is the number of personnel out looking for leaks and then repairing them. Leak location and repair is sometimes called active leakage control (ALC).

There is a steady state situation in which the marginal cost of the ALC effort is equal to the marginal cost of the water saved by adopting that ALC policy. This is the short-term ELL.

Long-term ELL

In the longer term, investment in facilities such as district metering, telemetry, pressure management and mains renewal will have an impact on the short term ELL. The reduction in the short term ELL and consequently the savings and costs associated with the change can be compared to the investment cost of making the change.

Investment costs are sometimes called transitional costs, i.e. they represent the cost of making the transition from one steady state to another. Investments (sometimes referred to as 'intervention options') can be for leakage management, demand management, or water resource development.

The short-term ELL is based on an economic analysis which estimates the optimum level of ALC effort taking account of the costs of ALC and the short-term value of the water in the supply zone. Long-term ELL is based on a form of investment analysis taking account of the following questions:

- What is the current level of leakage?
- What is the short term-ELL?
- How will the short-term ELL change with the investment under consideration?
- What is the saving in water losses and the change in ALC resources from the proposed investment compared with the current policy?
- What is the cost of the proposed investment?
- What is the return on the investment?

The answers to these questions will allow the water supplier to decide an investment policy using normal investment decision criteria.

4.3.3   Calculation of ELL

There are several methods of assessing ELL. The aim here is to set out the key factors, which must be assessed whenever the calculation of ELL is being considered for a water supply zone:

- *Cost of water.* The cost of water will include operating costs such as the power and chemicals required to treat and distribute the water around the network. It will also include longer-term investment costs. If reducing leakage will offset or delay the need for a new treatment works, then the deferment of that capital expenditure has a value. Similarly, reducing leakage by sufficient to allow closure of a water treatment work, or a change to network operation, will have a measurable value.
- *Short-term costs of leakage reduction.* These costs are usually limited to the cost of ALC. They include the costs of employing the staff to locate leaks, their vehicles, fuel and equipment.

- *Repair costs.* There is an argument which states that repair costs should be excluded from the evaluation of ELL. This is because the number of bursts which occur in any year is deemed not to change. The change in ALC effort will affect only the time (on average) for which bursts and leaks run before they are found. It will not affect the number, and so the number which have to be repaired in any steady state year will not change. Whilst this argument can be supported for steady-state situations great care must be taken when leakage is being reduced from one level to another. There will be a tendency to find leaks which would not otherwise have been found were it not for the extra ALC effort, and therefore the number of repairs being carried out will increase. Also, some of what is regarded as background leakage will be found by intensive ALC using new technology.
- *Long-term costs.* These include the net present value of the investment which is planned for leakage reduction measures, such as district metering, pressure management and mains renewal, over say a 20 or 30 year time horizon. Such intervention options have a one-off cost to reduce leakage to a lower level

4.3.4   ELL as an element of the supply–demand balance

Whenever possible, leakage should be considered as a part of the overall demand in an area, and not as a separate element. The calculation of ELL must then take into consideration the cost benefit analysis of leakage reduction in comparison to that of other options for maintaining the required level of headroom on the supply demand balance. The options available include:

Supply side options:
- building new sources and treatment works, or expanding existing ones
- laying new pipelines to bring water in from adjacent zones with more plentiful supplies
- buying water from neighbouring water supply organisations

Demand side options:
- controlling customer demand by new tariff structures
- metering customers who currently receive an open supply
- reducing customer use by installing water saving devices and appliances
- conducting water audits to minimise waste and undue consumption

All forms of leakage management are demand side options.

The options available should be compared in terms of their long run marginal cost, i.e. the costs analysed over a 20 to 30 year time horizon in relation to the benefit gained. The ratio of NPV of costs to NPV of the benefit expressed in units of cost/m^3 can be used to compare the available schemes. Schemes will give different degrees of benefit, and so the analysis should be conducted on an incremental basis, choosing the most economic measure for each increment (say in terms of Ml/d), and then reassessing the next increment on the basis that that measure is fully in place.

If such a long-term analysis is carried out, then it will show which leakage management measures are cost effective, and also the resulting long term ELL. It will also indicate the level of ALC effort required at each stage, and will enable a long term plan for leakage management in the zone to be derived.

Figure 4.6 shows how leakage is reduced, or available supply is increased by phasing in different types of demand and resource management scheme, in order to meet the growing need for water in an area.

4.4   SETTING TARGETS FOR LEAKAGE

Setting leakage targets goes beyond the stage of calculating ELL. Targets must be specific to a particular supply zone, and the overall target for the water supplier should then be the aggregation of each zonal target. Different procedures based on the analysis of ELL will apply in each zone, but there are also some policies which can only be applied at an organisational level. For example, it is difficult to have different policies for customer supply pipe leakage, or for internal plumbing losses, between different areas of the same organisation. Therefore, these global policies will influence the overall target compared to the true ELL.

The other issue is the influence of external pressures, which can significantly affect the target for leakage compared to the ELL. External factors include:

• Comparisons with similar water suppliers. It is inevitable that water suppliers will compare their current leakage level and their calculated ELL with that of

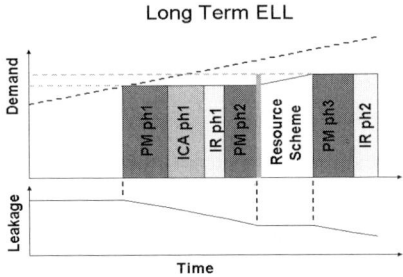

PM   – Pressure Management
ICA   – Instrumentation Control and Automation
IR   – Infrastructure Replacement
Ph1   – phase 1
Ph2   – phase 2

Figure 4.6 The impact of demand management and resource schemes on the long-term ELL.

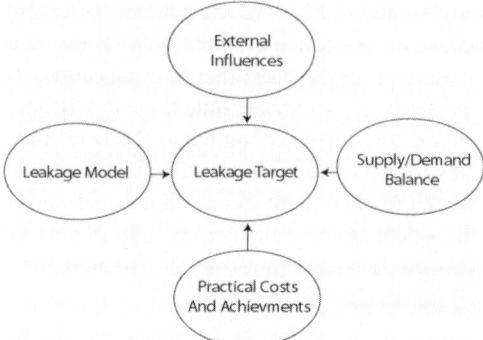

Figure 4.7 Factors influencing the target for leakage and losses.

other water suppliers working in the same country or within the same geographical region of that country.
- International comparisons.
- Political influences. The true ELL for a company with plentiful supplies and a poor infrastructure may equate for instance to 35% of total demand. However, the company may come under pressure from its customers or from government or peer pressure to reduce leakage because it is perceived to be too high.

In addition, the target will depend on the purpose for which it is being used:

- Is it a general target for the company as a whole, based on company-wide data, or is it an amalgamation of area and zonal targets?
- Is it a target set with reference to previous internal experience of leakage management over a number of years, or is it an initial target using a wealth of default data and comparisons with other organisations?

There is also the question of how the target relates to the true ELL, and indeed whether there can actually be a truly economic ELL given these external pressures. The calculated ELL will vary with time due to a number of reasons:

- Weather effects may result on short-term shortages of supply, leading to higher costs or values for water and so driving down the ELL in a particular year. For example, in a drought year, there may be the need to supplement supplies by imports from neighbouring regions or to use higher-cost supplies which otherwise are dormant. Once the drought is over, the ELL will rise to its longer-term average.
- Different weather effects may result in a spate of bursts and leaks due to ground movement, or frost damage. In these circumstances, the seasonal level of leakage may exceed the ELL, and it is not economic to provide additional resources to maintain leakage at the longer term ELL.
- The economics of supply and demand can affect the costs of leakage

management. In some countries there is a shortage of experienced staff, and so initial costs (demand exceeding supply) may be high resulting in higher profit margins for contractors. However, as the volume of work increases in a region, and more staff gain the necessary experience, competition between suppliers results in prices being driven down. This will tend to reduce the ELL

- New technology will tend to make leakage management operations more efficient, and once these techniques are well established they will tend to reduce the ELL.

Therefore the theoretical ELL will vary between one region and another and it will vary over time. However, leakage targets are less sensitive to such changes. It is unlikely that targets will be allowed to rise, even in unusual short-term situations. They are more likely to follow a downward path with an effective ratchet mechanism which prevents targets from drifting upwards. The aim should be to reduce leakage to a long-term optimum level, which is sustainable. Variations about this long-term average should be excepted and accepted.

4.5 THE IMPACT OF REGULATION ON THE WATER INDUSTRY

The most significant external factor affecting leakage targets is the impact of regulation on the water supplier. The nature of the regulation will depend on the general control and ownership of the water supply organisation in a particular country. In the UK, for example, the water companies of England and Wales are privately owned, and are subject to stringent regulation. All the privately owned companies are required to submit an annual return to the Office of Water Services (OFWAT). In Scotland, where the water authority is in public ownership, the Water Industry Commissioner also requires annual returns, and in Northern Ireland (where the water supply system is also publicly owned and operated) the leakage management plans are subject to scrutiny by the government Audit Office.

Another example is the voluntary system which operates in the USA, under which some utilities send input and consumption volumes on a voluntary un-audited basis to American Waterworks Association (AWWA).

Regulation creates expectations from several parties which have to be managed:

- Customers expect the water supplier to control the prices they pay for their supply. Too much leakage is wasteful, but so too is expenditure on leakage management which does not give an adequate return.
- Government economic regulators (such as OFWAT in the UK) also expect operating and investment costs to be justified.
- Shareholders expect the water supplier to be managed efficiently and to produce a return on investment.
- Environmental regulators and pressure groups seek to avoid further abstractions of raw water which deplete streams, lakes and rivers, and there are often protests against the building of dams.

- Drinking water inspectors protect supplies, and seek to control any works which may have an adverse effect on water quality. Leaking pipes do not present a good image for a company focused on providing a quality service. Leakage control works come under scrutiny from water quality regulators due to the disturbance they may cause (e.g. step testing) which can result in discoloured water.
- National governments aim to manage water supplies generally to ensure there is a general sufficiency for public health and for economic development.

4.6   A PRACTICAL APPROACH

It is difficult to be prescriptive as to the methodology which should be followed for setting leakage targets and for deriving an appropriate strategy for managing leakage. In some countries there will be little guidance given to water suppliers to follow a particular method. In others, the nature of the water supply industry may preclude significant investment in leakage management and the time required to formulate a strategic plan – for example if there are many hundreds of small water supply organisations, each working under different regulations. In these cases, it is often left to the water distribution manager, who has responsibilities for several functions, of which leakage is only one, to guide the strategy and also to implement the actions.

The following sections (4.6.1 to 4.6.10) set out a general series of steps to follow, whatever the nature of the organisation. In small organisations these steps may form the overall business plan for leakage. In larger organisations, the guidance may be useful to the leakage manager, who will be responsible for only part of the area of supply. He/she is not likely to be involved in setting global targets, but will need to understand how targets should be set, and who will want to know whether the targets he/she has been set are achievable for the operating environment.

Figure 4.8 illustrates the generalised leakage management strategy.

4.6.1   Understanding the factors which drive down leakage

The prime questions are:

- Is there any external influence on leakage which requires the supplier to make a reduction or are targets set internally?
- If there is an external influence on leakage, are there mandatory targets, or is any influence optional?

Whilst targets may not be mandatory, there may be a requirement for leakage data to be submitted annually or at other regular intervals. There may be a requirement to submit other data such as the extent of pressure management, and the policies and procedures which the water supplier has adopted to control leakage.

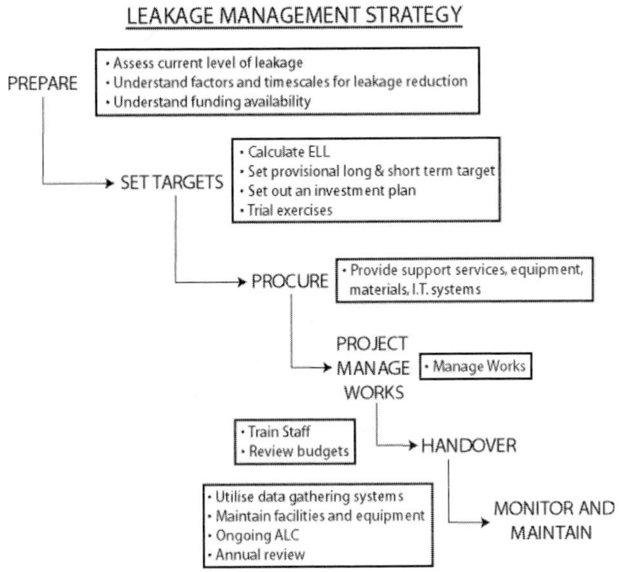

LEAKAGE MANAGEMENT STRATEGY

Figure 4.8 General leakage management strategy.

This data may be used as part of an overall service assessment, with some mechanism in place for penalising poor performance.

If targets are set internally, without any external influence, there is a need to understand why targets have been set, i.e.

- Is there a current water shortage generally, or in a particular region, or is a shortage forecast in the near future? There are cases where the treatment works supplying a town may be reaching the end of its useful life and water quality does not meet new standards, so the works has to be replaced or refurbished. In order to control capital expenditure, there will be a reluctance to size the new works at such a level that it is capable of meeting not only the forecast customer demand but also the current leakage level if it is considered to be in excess of a reasonable amount. In these cases, the target for leakage will be set in conjunction with the capacity requirement for the new works, such that the future supply demand balance in the zone has an acceptable margin of safety.
- Have targets been set to bring the supplier into line with other similar organisations, or has the country as a whole become embarrassed about high leakage levels making it a political issue? In this case the answer is to understand the time-scale for leakage reduction. Whether targets have been set on external or internal influences, they will have a time-scale attached to them. A general target may be to reduce leakage by 20% within 5 years, or it may be to achieve

a value for leakage in litres per property per day by a given year, that year having some other significance. Time-scales must be realistic.
• Are there internal savings to be made which are sufficient to justify a programme of leakage reduction?

There is no easy solution to controlling leakage, even if substantial finance is made available for a one-off exercise within a constrained timescale. Leakage management is a painstaking process, and projects which aim to make a substantial reduction in leakage in a short space of time, are unlikely to succeed in the long term. They often use data and assumptions based on experiences in other places, which are not translated into the local environment. They rarely include provisions for training and technology transfer to ensure that leakage is maintained at a lower level by local staff once the finance comes to an end.

Attempting to achieve too much too soon will lead to inefficiencies. There is a natural pace to a leakage reduction plan, which varies from one region to another depending on a number of different parameters.

4.6.2   Setting a provisional long-term target

Where no particular internal or external influence exists and there is simply a general desire to reduce leakage down to an economic level, then it is recommended that a provisional long-term target should be established as a goal to work towards. This goal should be ambitious, but it should not be impossible to achieve. The target should not be cast in stone. The water supplier should be prepared to change the target as new data becomes available. He should be satisfied if the target has to be revised upwards, and if the initial target is achieved more easily than expected, he should be prepared to consider a new lower target if it can be cost justified.

4.6.3   Setting a short-term target

The short-term target should be set with reference to the long-term target. A reasonable approach is to aim to achieve between 50% and 80% of the long-term reduction within say 5 years. This time-scale is chosen as a reasonable one in which to set up the necessary facilities, to award contracts, and to carry out the initial works.

A time-scale of between 4 and 7 years is reasonable; any less is too ambitious, and any more will not be as economic. The costs of leakage reduction will still be the same, but there is the cost of supplying the excess leakage after year 7.

Leakage will tend to be reduced over time following the 'S' curve shape shown in Figure 4.9. The early leakage management expenditure may show little return on the investment, and there is a risk that staff will lose faith. However, if the effort is maintained (usually after 1 to 2 years), then leakage will begin to

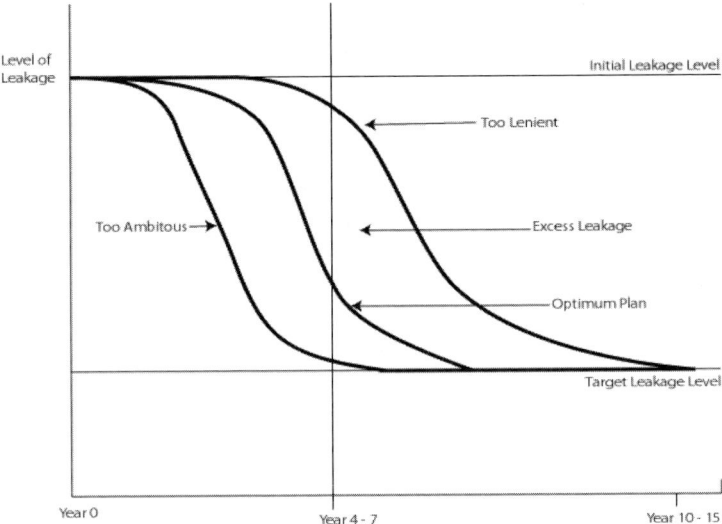

Figure 4.9 Optimising the timescale for reducing leakage.

reduce quickly giving significant reductions in years 3 and 4. In the final stages, the rate of reduction will slow down due to the laws of diminishing returns explained earlier.

One UK leakage manager found that resource and organisational constraints made it difficult to achieve anything more than a 15% reduction in leakage in any year. This factor tended to prevail over the calculation of the long-term ELL, when setting annual targets.

4.6.4 Setting up procedures to collect data

Before any major expenditure is made on leakage reduction, it is very important that procedures should be established to collect all relevant data. These data will fall into three general categories, operational, tactical, and strategic – these are expanded in Figure 4.10.

Data will be collected at a micro level (usually by DMA) and at a macro level (based on the general water balance). Micro level data will be subject to ongoing review resulting from data collected from leak detection exercises, or from studies into pressure management and district metering. Macro level data will generally be updated annually. More external data sources, such as government statistics on population, will be used. It will also be subject to greater external scrutiny – often any changes to the base data have to be justified to external parties.

OPERATIONAL DATA	Current leakage levels
	Current pressure data
	Outlet settings of PRVs
	Records of the number and type of leaks found
	Records of hours spent on ALC
	Industrial water use
TACTICAL DATA	Zone boundaries
	Types of PRV
	Maintenance records
	PRV performance records
	Asset data
STRATEGIC DATA	Distribution input averages
	Water balance calculations
	Results of studies and pilot exercises
	Lessons learnt database
	Numbers of ALC staff employed

Figure 4.10 Data collection for leakage management.

It is important that some attempt is made to reconcile estimates of leakage based on top–down and bottom–up methods. The aggregation of DMA losses to give an estimate of total company losses can sometimes be misleading for the following reasons:

• Leakage on trunk mains and service reservoirs outside DMAs will have to be estimated and added to the total.

• There may be issues surrounding the timing of DMA meter readings, and the level of operability (i.e. the number of DMAs which are working correctly at any particular time, with reliable meter readings).

Data collection should be established in a hierarchical way, so that by aggregating the data for DMAs within a supply zone, it is possible to provide average data for that zone. Similarly, supply zone data should aggregate to the company average values. In some cases it is necessary to create un-metered DMAs – these are areas which are not actually covered by a district meter, but for which the water supplier does have relevant data (e.g. number of properties and length of mains in the area) to make an estimate of leakage by other means. They usually have a defined boundary.

Leakage management is data hungry, and so investment in software systems and staff resources to manage the data can be significant. However, without such systems, there is a risk that investments will be targeted ineffectively, and leak detection and repair operations will be carried out inefficiently. Therefore, although the cost of establishing good data management systems represents a high initial outlay, there will be a return on that investment in the longer run.

It is worth noting that the study of leakage statistics is not a precise science. Although it can be deemed similar to financial accounting in nature (inputs compared to outputs in a balance sheet), it is very different in terms of the level of accuracy. With leakage statistics, there will always be a number of unknown values which have to be estimated or averaged.

4.6.5 Undertaking an annual review to assess the effectiveness of the strategy

It is recommended that the leakage management strategy be kept under constant review, and that it is subject to some form of annual audit. This could mean a review by senior management, or it could involve external consultants. The review should include the following elements:

- progress against target
- changes to target due to lessons learnt
- changes to assumptions and default data
- investment made

4.6.6 Reviewing the organisation structure and making changes to leakage programme management

Where a major leakage reduction programme is planned, it will be not be efficient, nor is it likely to be fully effective, unless the organisation structure is reviewed, and changed to take account of the new demands placed upon it. Whenever the workload warrants the appointment of a leakage manager, this is the preferred course of action. The leakage manager should then have responsibility for delivering the leakage reduction programme along the strategy agreed by the directors. It is essential that the leakage manager be given the support of the directors, and that if possible the same person remains in post during the leakage reduction stage. That person, who should be committed to the project, will become a focal point for leakage management activity and will be able to coordinate the various aspects of the programme.

There will inevitably be times when the programme is not going according to plan. During such periods it is important to differentiate between two stages of leakage management:

- Leakage reduction down towards a set target. This stage should be regarded as a project involving capital works, and other transitional costs, which can be managed as a project in a similar way to building a new treatment works. The project should be managed by someone with good project management skills, who does not necessarily have to be a leakage expert. However, the team should incorporate experts or external advisers, and also involve operational staff who will be responsible for maintaining leakage once it is reduced.

- Maintaining leakage at the target level. This stage should be incorporated into the ongoing management of the water supply organisation, and managed in a similar way to the operation of the treatment works. During this stage there is less need for a dedicated leakage manager, but if the organisation is sufficiently large there is a case for a leakage coordinator to ensure consistency of approach across the region.

These two stages may run concurrently in the same organisation. A supply zone will be handed over from operational staff to the project management staff, and once the leakage reduction work is complete it will be handed back. Different zones will be at different stages during the overall project until work has been completed across the region.

4.6.7   Understanding the starting position

An essential element of the leakage strategy is the understanding of the source and level of leakage.

Strategies to get a better estimate of actual losses include:

- distribution input (DI) verification – reconcile leakage in DMAs with losses in the supply zone
- per capita consumption (PCC) studies
- property counts
- non-household water use
- operational use

4.6.8   Establishing trial exercises

These demonstrate benefits in a small area before introducing them company-wide. Whenever possible, demonstration exercises should be undertaken to show the techniques and also to deliver some early benefits in the leakage reduction programme. The exercises should not be regarded as separate from the mainstream programme. They should be planned as an integral part of it and carried out by the same team which will deliver the main body of work.

Pilot exercises may focus on a particular aspect of the leakage management programme, e.g. pressure control, or they may be all embracing to test the integrated strategy in a limited geographic area, e.g. a particular supply zone.

4.6.9   Establishing funding requirements

An appropriate leakage reduction plan can only be set with due regard to the funding requirements. Even if leakage targets can be shown to be economic, the work has to be funded up front in order to achieve a pay-back over a longer term – in some cases over 20 years. Funding could come from raising charges to customers, from government sources, from international grants and loans, or by accepting a lower level of profit during the leakage reduction work.

Another aspect is the impact of leakage reduction on income to the water supplier. If all customers are metered and leakage reduction involves reducing leakage on customer service pipes then there could be a requirement to review tariffs as part of the leakage management strategy. Some water supply organisations receive income from the municipality based on the volume of water they supply from their treatment works through their mains networks. Others are paid a concession to manage the municipal supply and distribution system, and they are paid according to customer demand. In these cases, the impact on income has to be assessed before any leakage reduction programme can be justified.

For these reasons, it is important for the sponsors of the leakage programme to involve the water supplier's financial managers at a very early stage, as resolving the funding issues could take just as much time and effort as the technical aspects of the strategy

4.6.10   Highlighting the savings from leakage management

The savings which can be made from leakage management are set out in Appendix A. These savings are varied and include operational costs and investment costs. They include both short- and long-term savings. They often lead to savings in several budget areas, including budgets which are controlled by personnel with little or no responsibility for the leakage management programme. For example, a reduction in leakage will reduce the throughput of a treatment works, leading to savings in power and chemical costs. It is unlikely that the manager responsible for the treatment works budget will be the same person responsible for the leakage programme. Therefore, unless the savings are forced out of the budget there is a risk that the savings will not actually materialise and the budget underspend will be targeted towards other lower priority expenditure at the treatment works.

For this reason, the impact of the leakage management programme in terms of both expenditure and savings requires a high-level review of expenditure plans at the outset.

4.7   COMPUTER MODELS TO ASSESS LOSSES

Computer models have been used for many years for a number of water supply and distribution purposes. Network analysis is used to model the flow of water through the pipe system and to calculate flows and pressures at key nodal points. Water quality parameters can be modelled from source to tap to predict concentrations of chemicals. Computer models are also used to forecast demands, and to model inflows and demands on storage reservoirs.

Until quite recently no such techniques were available to model leakage levels. This was due to several factors:

- a lack of understanding about the mechanics of leakage paths
- an empirical knowledge only of the relationship between leakage and pressure
- consideration of leakage as a single element without understanding its individual components

The National Leakage Initiative, which was carried out in the UK between 1990 and 1994 [5], identified a need for a method of collating the information arising out of a number of sub-groups which considered several aspects of leakage management. The outcome was a general methodology which came to be known as BABE – Burst and Background Estimates [6]

BABE modelling is not a precise science. It relies on a number of estimates (as the name implies) and assumptions some of which are based on good quality data specific to the water supplier. Some are default values based on industry averages, and others are based on sound engineering judgment. The relative balance of these estimates will affect the overall accuracy of the results and the reliability of the model which is created. However, that should not detract from the technique. In principle there is little difference between it and the general concept of pipeline network modelling, which also uses a mix of measured data from field tests, and estimated values.

The objective of BABE modelling is to assess the individual components of leakage in a supply zone, and then to compare that estimate with the level of leakage derived from either the water balance or from nightline data, or preferably both.

In 1994, after the results of the UK Leakage Initiative were published, a single BABE computer model was launched and a user group was established to compare the results obtained by a number of UK water companies. Since then the same techniques have been applied internationally, and they have been developed and expanded to suit different local operating environments. Within the UK also, the original BABE model has led to the development of many other models using the same techniques. Some are very simple single page spreadsheets, others are comprised of several sheets for each supply zone. Some are linked to water demand forecasting models to give a least cost strategy for managing the supply demand balance over a 30-year timescale. In many cases, the models have a tool for calculating the economic level of leakage, and for carrying out 'what if' types of analysis.

4.7.1   BABE concepts

To understand how the BABE type models work, an explanation of the general BABE concepts is offered in the following sections. Leakage and losses are split into several components as follows:

Background leakage

Background leakage comprises numerous leaks which individually are quite small, but which can account for a significant proportion of total leakage because they run for long periods of time. In fact, it is thought that many background leaks are never repaired, and so they continue to run from the day they first occur until that component of the network is eventually replaced. They occur at pipe joints, fittings such as service connections, stop taps, or on the packing glands of line valves. In a well-managed system in which the leakage due to bursts is well under control, the background leakage is often found to account for the majority of total leakage.

The nature of the pipe material will have an impact on the level of background leakage. Welded polyethylene systems will be less prone to background leakage than metal systems with numerous mechanical joints. Plastic systems are not prone to small corrosion holes which occur in unprotected metallic pipe systems. Leaks are usually found by various methods of sounding. Leaks tend to make less noise in plastic systems, and so the level of background leakage may tend to be higher than in metal systems for this reason too.

Background leakage occurs on trunk mains, distribution mains and service pipes. There is a level of leakage from service reservoirs which can also be considered as background leakage because, although it can be measured, it may not be worth remedying by costly repair works.

Reported bursts

Bursts and leaks which come to the attention of the water supplier without the need to actively look for them, are known as 'reported bursts'. These may be leaks which come to the surface and so are visible to the passing public. Alternatively, they may impact on customer supplies and so they result from customer complaints. They tend to have higher flow rates than background leaks (though this is not universally true), but they are reported soon after they occur. They will also be repaired relatively quickly in order to restore supplies, or to make safe a highway which has been damaged by the high volume of water. Therefore, although reported bursts have a greater profile when water leakage comes under scrutiny, the total volume lost from reported bursts is usually relatively small.

Unreported bursts

These generally have flow rates lower than reported bursts, but higher than background leaks. They are only found by undertaking active leakage control (ALC) operations. They may run for only a few days, or they can run for several years, depending on the nature and intensity of leakage control activities. If no ALC is practised, and the water supplier only responds to reported leaks, this is known as a passive or reactive leakage control strategy. Where water is cheap and plentiful and there is no external pressure from government regulators to reduce leakage, then this may be an appropriate and economic strategy.

In any system the number and relative importance of these three types of leaks will differ, and there is no 'normal' situation. Factors such as ground conditions, soil type and whether the surface is developed or agricultural, the pipe material, and the pressure will all affect the proportion of reported to unreported bursts.

The importance of time

Whereas background leakage runs continuously, the level of leakage due to both reported and unreported bursts will depend on the time for which they run. The run time comprises three elements:

AWARENESS TIME

This is the time taken for the water supplier to become aware that the burst exists. For reported bursts, there is little or no control over this time and the only action is to ensure that members of the public can make contact without delay to report an incident. It is a useful element of leakage management strategy for all vehicles and press adverts etc. to carry a telephone number (sometimes called a leak line) dedicated for reporting leaks.

For unreported bursts, the awareness time will depend on the systems in place for collecting data on leakage levels. If meters are read monthly, then on average a burst will run for two weeks before it is known to the leakage manager. If data loggers are attached to district meters and these are downloaded weekly, then the average awareness time will be 3 to 4 days. If some form of telemetry is attached to meters, then the awareness time could be less than 24 hours.

LOCATION TIME

This is the time taken to locate the burst once the water supplier is aware of its existence. The key factor here is the number of staff available to carry out the active leakage control effort. The location time will also depend on the technology which is employed, and also the capabilities and efficiency of the leak location staff. However, for any given level of competence, there is a simple inverse relationship which applies. Doubling the number of staff will effectively halve the average location time, and hence the volume of water which is lost from unreported bursts.

REPAIR TIME

This is the time taken to effect a repair or shut off the burst once the location has been pin-pointed.

In theory, no more effort is required to effect a repair within two days, compared to doing it within 20 days. However, in practice a quicker repair policy may require staff to be placed on standby, and it is difficult to even out seasonal peaks and troughs in burst frequency. There is a balance to be made between a policy which is uneconomic because the time taken is too long, and water is being lost unnecessarily, and one which is too tight in which the additional cost is not offset by sufficient additional reduction in leakage level. In the UK unreported mains bursts and bursts on supply pipes in the public highway are usually repaired in 10–14 days.

For customer supply pipe bursts, and for mains on private land, there will be an additional time taken to serve appropriate notices, in order to gain lawful access to carry out the repair. There may also be a need to persuade the customer to effect the repair at his own cost depending on the legislation which applies. For these reasons, leaks and bursts on customer supply pipes can run for several weeks, and even months, before they are repaired. Consequently, although the flow rate from any burst may be relatively small, the total volume of leakage on customer supply pipes is usually a significant proportion of the total leakage. This is demonstrated in Figure 4.11.

4.7.2   Uses of BABE models

BABE-type models can be used for a number of purposes:

- To estimate the current level of leakage, and to aid the understanding of its components
- To assess the impact of different forms of investment such as pressure management, a change to ALC effort, infrastructure improvements, use of telemetry
- To estimate the short-run and long-run economic levels of leakage
- To estimate the background level of leakage in a DMA to assist with target setting

The models can be applied at a macro level across a whole region, using a single model for the whole organisation. This ensures that the modelled level of leakage

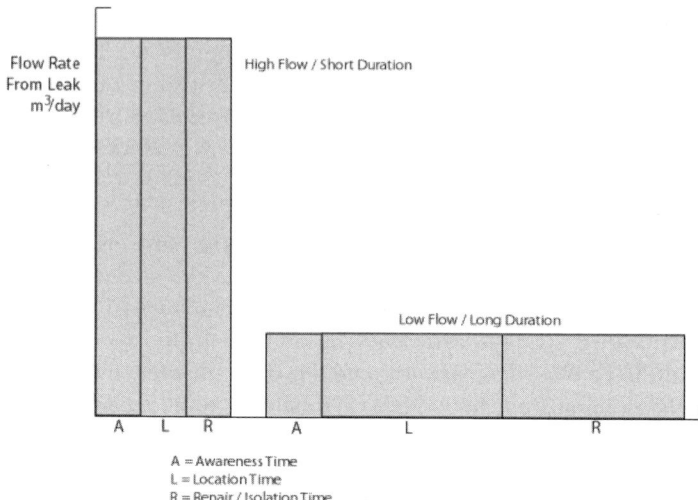

Figure 4.11 The influence of leak run time on leakage levels.

corresponds with the company estimate. It ensures that average values for mains bursts, lengths of mains etc, are reasonably accurate.

It also ensures that default averages are reasonable. However, it is difficult to estimate average pressures, or to assess the impact of particular schemes. At the other extreme, the principles can be applied to an individual DMA. Then the data on flows and pressures is more specific, but the number of mains and service bursts is more difficult to predict due to the variability in small areas. The ALC effort, which affects location time, will also be more variable at a DMA level. To obtain the best balance, BABE modelling is generally best applied to a supply zone of between say 10,000 and 50,000 properties. This level allows the BABE estimates to be compared to the level of leakage assessed from the water balance for that zone. It also ensures that data is accurately allocated, and it evens out variability in failure rates which drive leakage upwards.

4.7.3   Alternatives to BABE

An alternative to BABE is to take the level of leakage in an area as a single value and to then predict how that level will respond to different forms of investment, by using empirical information and rules. The advantage of this approach is that it is seen to use actual data specific to the particular organisation, and therefore there is a perception of increased accuracy. However, care should be taken as past results are not always the best predictor of future events. A comparison can be made to forecasts of water demand. In the UK in the 1970s water demand forecasts were based on past growth. Demand was projected forward according to some mathematical formula governed by the shape of the growth curve. Maximum and minimum limits were set in a seemingly arbitrary manner. The basic assumption was that demand would continue to grow due to similar factors which had occurred in the past. However, in the 1980s there were major changes in the UK which affected the demand for water. The coal and shipbuilding industries shrank to unthinkable levels. This reduction in demand was countered by growth in domestic use due to greater appliance ownership (washing machines, dishwashers), refurbishment of older properties to install additional bathrooms and WCs, and the reduction in the household occupation leading to greater use per person. However, the overall effect was a much reduced rate of growth from that which was forecast.

4.7.4   Allocation of leakage in network modelling

Network modelling provides a useful source of data for the leakage manager, and it is useful to combine these two activities in some way. Basic data on numbers of properties, lengths of mains, flows and pressures are common to both activities. Information relating to demand profiles from industrial customers also helps the analysis of night flow data.

A better understanding of the nature of leakage in a zone will improve the accuracy of the network model. Some modelling procedures distribute leakage evenly across all nodes. However, a better technique is to allocate leakage according to pressure, and to where bursts are most likely to occur. With the benefit of all mains models, it is sometimes possible to identify major leaks by analysis of the model calibration data compared to the modelled results.

It is recommended that integrated zonal studies be carried out, with leakage as one element of the analysis. The integrated approach to network management ensures that a common set of data and assumptions is used, and that expenditure plans are consistent. The approach avoids missing or double counting the benefits from different forms of investment on the distribution network.

4.7.5   Importance of company specific data

Whichever method is used to model leakage, or to derive a strategy, there is an inevitable dilemma. The strategy will be most important when the water supplier first embarks on a major leakage reduction exercise. To be accurate and reliable, the leakage model and the investment plan should be based on good quality data specific to that organisation, and a well thought out investment programme in which there is a degree of confidence that the expenditure planned will give the required benefits. However, if little or no work has been undertaken, then there will be no specific data. No data means that use has to be made of default values and assumptions and so there is less confidence in the modelled results.

Leakage modelling has been carried out over a number of years now, and the BABE concepts have been used in several countries around the world, with very different operating environments, and different policies, levels of service and configurations of their water distribution networks. As such, the original UK data have been enhanced, and the effect of these different global situations has resulted in a wealth of new data. For example, it is possible to model the level of leakage in systems which only supply water intermittently, for a number of hours a day.

The solution is to take gradual steps, using data from each step to make changes to the model and to the strategy based on the lessons learnt. It is almost impossible to set out an effective leakage reduction plan and to stay with it from start to finish (see Figure 4.12).

4.7.6   Infrastructure condition

The most significant factor affecting the level of leakage in a water network is the general condition of the mains and service pipes, and the service reservoirs. It is usually found that this is also the singularly most significant factor affecting the economic level of leakage for that network. The condition of the infrastructure is something that is inherited from previous regimes and generations, and it cannot be improved significantly without major capital investment in renewal and refurbishment works. It has been shown that expenditure on infrastructure improvement, even if it is targeted to the areas most prone to high leakage, is not a particularly cost effective method of managing leakage. Where improvements are being made for other purposes, such as the need to meet water quality parameters,

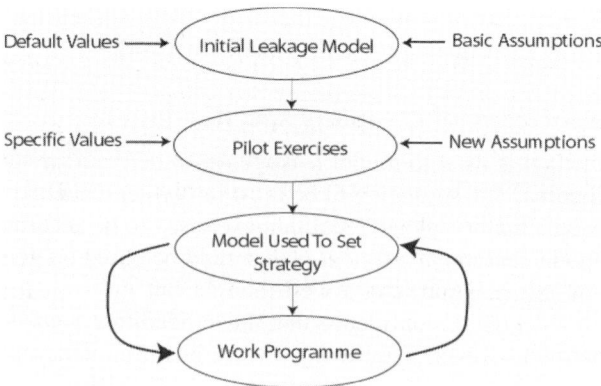

Figure 4.12 Using specific data to replace default values and assumptions.

or customer standards of service for interruptions to supply or minimum pressure standards, then the impact on leakage levels should be taken into account. However, where leakage is the primary problem then it is difficult to cost justify mains renewal works.

The condition of the infrastructure can be assessed in two ways:

- its propensity to burst – this will be affected by factors such as pressure and ground conditions and weather as well as the condition of the mains and service pipe fabric
- its propensity for background leakage – again this is affected by pressure.

At first sight it would seem that these two parameters would be linked, and that mains in poor condition would be susceptible to both effects. However, it does not follow that a high burst rate means high background leakage and vice versa. Therefore a separate judgment has to be taken on the two parameters.

Burst frequencies can be assessed by inspecting the records of repairs carried out over a number of years, to evaluate a typical average frequency. Care must be taken to relate this to the method of leakage control in place over the same period. If the method of control is to be changed, e.g. from passive to active, or from a low level of activity to a more intense level over a transitional period, then the frequency of leaks repaired in future will be different to that in the past.

The condition of the infrastructure in terms of its affect on the level of background leakage is referred to as the infrastructure condition factor (ICF). This can be assessed by considering the level of background leakage which can be achieved in a sample of DMAs (or all of them) after all unreported bursts have been found and fixed. This can then be compared to the level of background leakage for typical average values using BABE principles. The ICF is the ratio of the actual lowest achieved leakage rate in the DMA, to the rate predicted from the values contained in the *Managing Leakage* reports [5]. As the ICF of districts can vary considerably from 25% of the average value (as set out in *Managing Leakage* Report F) to in excess of 400% of average, the sample must be sufficiently large to be representative. It is typical for 25 to 50% of districts to be included in the sample, and ideally as many DMAs as possible should be included to improve the accuracy of the ICF estimate.

The variability of ICF values for the supply zones of one UK water company [7] is shown in Figure 4.13. When the ICF of each individual DMA is compared, the variation is much greater. Some of the higher DMA values (in excess of 3) may be due to bursts which have not been located, and some of the lower values may be due to inaccurate data. If the ICF of a DMA is less than 0.5 it is recommended that the data should be audited to confirm that it is correct. In some cases negative ICFs result from inaccurate flow data.

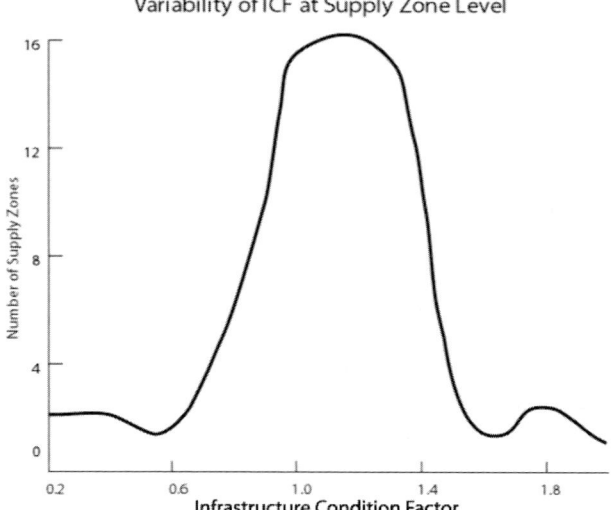

Figure 4.13   Variability of infrastructure condition between supply zones.

4.8   DESIGNING AND INTRODUCING A STRATEGY

There are several key stages to a leakage management strategy:

- understanding
- quantifying
- setting targets
- implementing
- monitoring and maintaining

There is no standard approach or a model strategy. The strategy must be tailored to each individual water supplier, and ideally to each supply zone to ensure it is appropriate. It must be:

- specific, to take account of all relevant factors, and avoid too many comparisons with other organisations
- timely
- integrated with other supply and distribution works
- comprised of different strategies to gain control of different sources of leakage i.e.:
  - supply pipe leakage
  - trunk main leakage
  - service reservoir leakage and losses
  - DMA leakage

- inclusive of issues such as:
  - training
  - equipment
  - communications (both internal and external)

4.8.1 Introducing the strategy

Perhaps the most important aspect of the leakage management strategy is that it is understood and followed by all parts of the water supply organisation. There is a need to gain the commitment and cooperation of staff in a number of different departments, and in regional offices and depots of a large diverse organisation. Effective leakage management requires an input from a number of different personnel, and unless they are all committed, the implementation of the programme will not be efficient, and it may then be difficult to maintain the infrastructure which has lead to the lower leakage levels.

For example, the introduction of district metering and pressure management requires line valves to be shut permanently in order to create discrete supply areas. In the event of a customer service problem, or a broken main, or during mains renewal and rehabilitation works, these valves may have to be opened temporarily. It is essential that the valves be shut again after the event has been dealt with or the works carried out. The issue of boundary valve management is one which has lead to the development of new technology, some simple and some quite complex. However, it is a problem which can and probably should be solved by ensuring that all staff with access to the network appreciate the importance of maintaining the integrity of the zone boundary.

As an integral part of the strategy, careful consideration should be given to:

- A launch event such as a major seminar. This helps to present the message that the programme is underway, and can be used to explain to staff how it may impact on their particular jobs, who is directly involved, etc.
- Education and training of all staff, not just those directly involved with delivering the leakage management programme. Use should be made of awareness exhibitions, seminars, internal and external training course, articles in-house magazines.
- A 'carrot and stick' approach to staff management, which has been found to be successful. Staff responsible for the ongoing operation of the water distribution system are rewarded for proper maintenance of the leakage management infrastructure, whereas disciplinary action is taken against those who wilfully neglect to take account of the need to keep these systems in place.
- Public relations. The public profile of the water supplier can be raised by a variety of PR exercises such as:

- good news stories in local newspapers
- technical articles in trade press
- advertisements
- providing information leaflets to customers

A useful technique is to establish a dedicated telephone number for members of the public to report leaks. The number can be publicised on the sides of vehicles, and in papers issued with customer accounts.

4.8.2   Role of consultants and contractors

It is unlikely that at the outset of a leakage reduction programme that the water supply organisation will employ sufficient staff with the necessary training and experience. The strategy should consider which elements of the work require external support, and the procedures to be adopted for procuring the services.

A long-term approach should be taken with contracts being let for a period of two years or more. Renewing contracts annually leads to inefficiencies due to start up and wind down. Therefore, care should be taken to ensure that the consultants and contractors can demonstrate a proven track record, and that they have the necessary resources available in the area to carry out the contract. Types of contract are discussed in 6.6.

4.9   REFERENCES

1   Parry, J (1881) *Water – Its Composition, Collection and Distribution*. London: Frederick Warne.
2   Gledhill, EGB (1957) 'An investigation of the incidence of underground leakage and an improved method of waste control', *Journal of the Institution of Water Engineers*, 11, 117.
3   Technical Group on Waste of Water (1985 [1980]) *Leakage Control Policy and Practice*, Standing Technical Committee Report no. 26. Original publication London: Doe/NWC. Reprinted London WAA/WRc.
4   Shore, DG (1988) 'Economic optimisation of distribution leakage control', *Journal of the Institution of Water and Environmental Management*, 2(5), 545–551.
5   WSA/WCA Engineering and Operations Committee (1994) *Managing Leakage: UK Water Industry Managing Leakage* Reports A–J: Report A – *Summary Report*; Report B – *Reporting Comparative Leakage Performance*; Report C – *Setting Economic Leakage Targets*; Report D – *Estimating Unmeasured Water Delivered*; Report E – *Interpreting Measured Night Flows*; Report F – *Using Night Flow Data*; Report G – *Managing Water Pressure*; Report H – *Dealing With Customers' Leakage*; Report J – *Techniques, Technology and Training*.London: WRc/WSA/WCA.
6   Lambert, AO (1997) 'Managing Leakage: Strategies for Quantifying, Controlling and Reducing Water Losses, based on Analysis of Components using BABE Concepts'.

Paper presented at IIR Conference – Water Pipelines and Network Management, London, February 1997.

7   Trow, SW (1997) 'Formulating an economic optimum plan for investment in leakage management: methodology and practical application'. Paper presented at IIR Conference – Water Pipelines and Network Management, London, February 1997.

5

Upgrading the network

5.1   INTRODUCTION

Many of the recommended economic and operational practices presented in previous chapters presuppose a developed infrastructure and established operational procedures.

Not all countries or regions have such a luxury – many are struggling to ensure that customers receive a reasonable water supply to sustain health and life, often in a network with an outdated infrastructure, with poor record systems, with inadequate technical skills and technology, an unsuitable tariff structure or revenue collection policy, and a poor operation and maintenance policy

The purpose of this chapter is to review the tasks required to update or modify the infrastructure to address such issues, and to implement the required changes.

5.2   THE ZONING CONCEPT

The principle of dividing the supply network into sub-networks or zones has been discussed in previous chapters. Traditional open systems, while maximising the use of the mains network and allowing free flow from all sources of supply, can cause a number of problems for the operator, with the network being subjected to mixing of water from sources of differing quality and pressure. An open system is also vulnerable to pump failure or loss of power supply, poor system information, and poor control. By dividing the network into smaller 'compartments' the water engineer can understand and more easily analyse pressure and flow profiles and problem areas. Such zones are operationally easier to manage, and allow monitoring and control systems to be implemented more easily.

Zones can be created as discrete pressure areas, e.g. a zone which cannot be supplied by gravity can be isolated by boundary valves and supplied by a booster pump. Zones can be created to separate water supplied from different sources, to minimise quality problems. The zoning hierarchy concept, and the creation of smaller zones, is described in Chapter 6 and is illustrated in Figure 6.1

5.2.1   Creating a zoned network

In any network, closing too many valves reduces the capacity of the network and may lead to operational or water quality problems. Therefore, wherever possible zone boundaries should be natural hydraulic and geographic boundaries, to minimise the number of mains crossing them. Examples of such boundaries are railways, rivers, canals, and main highways. There is usually some degree of redundancy in the mains network to allow zoning without causing supply problems. The ideal zone will have the following characteristics:

* single source, to minimise quality complaints
* single metered input, to maximise accuracy of flow metering and leakage data

  The sequence for creating zones is:

* identify possible zone boundaries using maps, mains records and site inspection
* carry out an audit of the proposed zone, including checking the status and condition of proposed boundary valves
* collate data on the zone characteristics (number of properties, industrial consumption etc.)
* measure zone flows and pressures by network analysis or field test
* isolate the zone and collect diurnal flow and pressure data using survey meters and pressure gauges/transducers
* calculate the required meter size
* evaluate the potential for pressure management
* install equipment
* mark boundary valves with a clear identifier

Table 5.1 Applications for different types of meter.

Principle	Type	Supply meter	Zonal meter	Customer (household) meter	Revenue (non-household) meter
Pressure differential	Venturi	√			
	Dall	√			
Mechanical	Volumetric semi-positive			√	√
	Turbine		√		√
	Inferential	√	√		√
	Combination		√		√
	Insertion		√		
Electro-magnetic	Full bore	√	√		√
	Insertion		√		
Ultrasonic		√	√		

5.3   FLOW METERING

It is vitally important to know accurately the load on the network. Continuous flow measurement at the source or reservoir outlets with data transmitted instantaneously to the operational centre is the ideal, but chart recorders or digital data loggers can be effective substitutes. Effective management of the network relies on the ability to monitor flows continuously, at a minimum of hourly intervals throughout the day.

An accurate population count is also of prime importance, as derived data such as per capita demand provides information on growth of demand over time, leakage etc.

Effective metering is an essential feature of network management, particularly for construction and calibration of a network model, and with flows into and out of each zone measured to provide an equation known as a 'water balance' or 'water audit'.

Table 5.1 indicates which type of meter is suitable for each level of the zoning hierarchy.

5.3.1   Meter accuracy

Meter accuracy depends on its type, age and condition, and water quality. Meter sizing is very important as the accuracy is expressed as a percentage of the flow range. If meters are oversized there will be an unnecessary loss of accuracy on the actual flowrate.

Venturi and Dall meters are rarely better than ±5%, and if installed in poor hydraulic situations, (usually because of insufficient straight pipe upstream and downstream of the meter), they can be very inaccurate.

Mechanical meters can be better than ±2% in good conditions, but at low flows, this will fall to around ±5%. Combination meters are accurate over a wide flow range because the smaller meter, measuring low flows, is designed to work within its optimum range.

Insertion meters are often not better than ±2–5% and can be very inaccurate if not used by trained experienced technicians. With the advent of electromagnetic insertion meters, accuracy has improved.

Modern in-line electromagnetic (EM) meters are accurate to ±0.1% of flow, but accuracy falls off rapidly at low flows. Older types suffer from drift and have to be re-calibrated at regular intervals.

Ultrasonic meters are improving, but they are not yet as accurate as electromagnetic meters. They have the advantage that, in the case of clamp-on meters, they can be installed without taking any pipe out of the line.

5.3.2   Installation requirements

The following installation criteria are recommended:

- Straight pipe lengths before and after the meter – the requirement differs for each type, but a standard of 10 diameters upstream and 5 diameters downstream is common practice.
- Valves must be installed either side to allow removal of the meter for maintenance. Some types can be maintained in the line by removing a flanged lid on which is mounted the working parts, and replacing temporarily with a blank lid.
- A by-pass must be installed to allow flow when the meter is removed for maintenance.
- A chamber is required for access to read the meter. In the smaller sizes, this can be small, extending only down to the top of the meter body, the rest of the meter being buried. In larger sizes, it is usual to build a chamber extending below the pipe invert. Modern electromagnetic and ultrasonic meters are designed to be buried, with the instrumentation mounted remotely, perhaps in a cabinet.
- Where access is difficult, routine reading can be achieved by installing a remote reading instrument ('out-reader').

5.3.3   Other options

Not all water undertakings have the capability to measure water into supply –
some rely on pump curves or historic estimates. Some alternatives are:

• Measure raw water (flume or meter) upstream of treatment works and subtract
  losses/use in treatment process.
• Measure downstream of treatment works (install bulk meter or temporary meter).
• Meter downstream of reservoirs (or reservoir drop test, but this will exclude
  trunk main losses).

Temporary meters are either insertion type, inserted through a pressure housing
or gate valve via a 2" (50 mm) BSP tapping. The electromagnetic type is more
accurate than the turbine type. Care is needed to ensure that at least 10, and preferably
50, straight pipe diameters upstream of the bulk meter and to avoid bends etc.
which affect the velocity profile. An alternative is the clamp on type ultrasonic
meter but this costs about 4–5 times as much as an insertion electromagnetic meter.

Water audits and water balance calculations, as described in Chapter 2, will
determine network metering requirements.

5.3.4   Procedure for installing a supply meter

Assess what supply data is available:

• assess the source of the data (i.e. measurements or estimates)
• decide validity of data
• verify data via a check meter, drop tests, or insertion meter
• install bulk meter or re-calibrate existing meter
• recalculate production figure
• recommend regular e.g. annual check using insertion meter or clamp-on ultrasonic

Large diameter electromagnetic meters are usually recommended for permanent
installations as they are relatively maintenance-free (except for an occasional check
on the secondary instrumentation).

5.4   ZONAL METERING

Zonal metering into zones of 10 000 to 50 000 properties is usually the first stage
of leakage monitoring. Zone size is usually defined by supply areas, pressure
boundaries, or natural hydraulic/geographic boundaries. Examination of total
integrated flow can pick out increases in demand or losses. Further zoning within

the supply zone is recommended for continual night flow monitoring in DMAs and for proactive leakage management.

The concept of zoning is more difficult to apply when working with practitioners in less-developed networks, particularly those with intermittent supplies. Some engineers insist on keeping an open system for maximum supply, perhaps operating valves daily to apportion supply at different times of the day. Some practitioners operate valves at night to reduce pressure, and thereby reducing mains failure frequencies.

The best way to demonstrate the value of zoning is to start with a pilot zone. To create a pilot zone in these circumstances, the actions described in section 5.2 should be followed, with these additional steps to take account of the nature of the supply:

- find and repair leaks as soon as possible
- observe the change in demand
- ensure that consumers' taps are closed (they are often left open to maximise collection of water when the supply is restored)
- use the results and the reduction in the intermittent nature of the supply to convince local engineers of the value of zoning
- repeat in other zones
- consider creating DMAs within zones

The most commonly used zone meter is the Woltman or helical vane inferential meter, which are widely available. These meters have a wide turn-down ratio (maximum to minimum flowrate) and can be used to accommodate peak demand as well as minimum night flows.

Strainers upstream are advisable as treatment processes in some networks allow debris through into distribution mains, which can damage the helical vane. However strainers require regular cleaning.

Maintenance is usually simple – replacement heads are easily exchanged in the field.

Calibration of these meters is not usually necessary nor cost-effective – they are usually replaced on failure. A regular inspection of the data is advised to verify that the meter is turning and the readings appear reasonable.

Electromagnetic meters which can be buried are being increasingly used for zone and DMA monitoring. No chamber is needed – the electronics are ducted to a read-out in an above ground kiosk. This could be a favourable point for developing countries.

5.4.1   Selecting a zone meter

State of the art flowmeter technology makes it possible to select a meter which can cope with peak daily flows and seasonal demand, and which can also accurately measure:

- night flows into a DMA
- night flows into sub-divisions of a DMA
- the very low flows associated with step testing

The choice of meter size and type will depend on:

- size of main
- flow range
- reverse flow requirements
- accuracy and repeatability
- data communication requirements
- cost of the meter
- cost of ownership and maintenance requirements
- company preference

The flow range and accuracy requirements of the meter will depend on the mode of use. In the past, meters used solely for leakage monitoring in DMAs required good repeatability – for monitoring trends in flow data – rather than absolute accuracy. However DMA meters are increasingly being used to gather total leakage data from the DMA system – a requirement of OFWAT in the UK – and accuracy of individual meters is an important requirement.

Accuracy is also important if a DMA contains several meters, measuring flows into and out of the district, as compounded errors in flow calculations could result in misleading leakage levels. In this case meter repeatability is also important.

The new generation of meters – mechanical, electromagnetic, and ultrasonic – are capable of measuring accurately over the flow range required for DMA monitoring. Mechanical meters, like the one shown in Figure 5.1, are most commonly used.

These have a turndown ratio of 100:1 or better, allowing flexibility in selecting the appropriate meter size. However, electromagnetic meters are being increasingly used, particularly where a full-bore meter is necessary to overcome poor water quality, debris in the main etc. The DMA market has resulted in cheaper, battery-powered versions of electromagnetic meters with a flow range comparable to mechanical meters.

Appendix B contains an example layout, and the schedule of materials required, for a typical zone/DMA meter installation

Figure 5.1 Mechanical meter installation, showing taper and straight pipe length.

5.4.2 Meter sizing

Rules for meter sizing should take account of headloss, seasonal fluctuation and demand changes. Where reverse flow has been encountered or is considered likely as a result of future operation a meter capable of measuring reverse flow will be required. Comparison of previous years' records will give an indication of seasonal differentials.

In areas with existing high leakage future flow characteristics, particularly minimum flows, can be significantly different. Allowance should also be made for the lower flows likely after leaks have been found and repaired.

If a network model is available, this can be used to predict the flow range of the DMA meter, taking into account peak and seasonal demands and minimum night flows. If a model is not available, a temporary insertion meter can be used to estimate the flow range, with some adjustment for seasonal and/or exceptional flows.

If neither of these tools is available the flow range can be estimated from demand calculations, using:

- number of properties
- estimates of per capita household consumption (pcc), estimates of metered use
- estimates of exceptional use > 500 l/hour (for maximum flows)

- estimates of night use (for minimum flows)
- estimates of leakage (for minimum flows after leak repair).

Maximum flow   = (no. households × pcc)
                    + (no. non-households × average consumption)
                    + exceptional use
                    + estimated distribution losses

Minimum flow   = (no. households × night use)
                    + (no. non-households × night use)
                    + estimated distribution losses after repair

The formulae used in the Burst and Background Estimates (BABE) software can also be used as a source of this information.

The flow ranges for a typical helix meters are shown in Table 5.2. The maximum flows quoted by the manufacturer are for exceptional flows for short periods – the figures for recommended continuous flow should be used for meter sizing.

In practice, the new generation of mechanical helix meters have such a high turndown ratio, that 'rule of thumb' meter sizing by properties/DMA is usually adequate, as shown in Table 5.3.

However, if headloss is an issue, resulting from the characteristics of the pipe network or bypass installation, a meter which is one size larger than those illustrated above should be considered, as long as it is still adequate at the lower end of its range for the DMA night flows.

Appendix B gives an example of installation design for a 100/150 mm meter on a bypass in a 150 mm main.

5.5   NETWORK RECORDS AND RECORDING SYSTEMS

Record-keeping is an essential part of water network management, and is also the basis for a GIS. Supply zone and DMA records should relate to both physical

Table. 5.2 Helix meter flow ranges

Meter size (mm)	80	100	150	200
Maximum flow (m³/h)	200	250	600	1000
Recommended continuous flow (m³/h)	120	180	450	700
Minimum flow (m³/h)	0.5	0.6	2.0	4.0

Table 5.3 Meter sizing per DMA size

Number of properties	Meter size
< 1000	80 mm
1000 – 1500	100 mm
> 1500	150 mm

records and records for leakage analysis. As well as PC-based records each DMA should have a dedicated paper-based file containing all DMA plans and records. Files should be kept in a DMA filing system, accessible to all leakage staff.

Appendix C lists the components of the network and the recording requirements relating to each.

5.6   SURVEYING THE NETWORK

If network records are poor, a network survey is essential before zoning and DMA design can take place, and for accurate leak detection and location to be carried out. There are several items of equipment which will support a survey of the network pipework:

5.6.1   Metallic pipe and cable locator

This is an essential pre-requisite for carrying out a pipe location and mapping survey prior to a leak detection survey (see Figure 5.2).

5.6.2   Non-metallic pipe locator

This is for locating PVCu and other plastic pipes (and asbestos cement/glass-fibre etc.). A mechanical oscillator is attached to a tap or hydrant, to induce a vibration in the water column. The signal is picked up by a receiver on the surface as it progresses along the pipe.

An alternative technique is to insert a flexible wire into one end of the pipe. A signal is induced in the wire, which is traced using a cable locator.

For newly laid non-metallic pipes consideration should be given to laying a metallic tracer tape along the pipe so that a metallic pipe locator can be used in future surveys.

5.6.3   Iron and steel box locator

This is a metal detector, used for detecting buried valve chambers and covers etc. during a pipe survey.

Figure 5.2 Metallic pipe locator.

5.7   PILOT STUDY AREAS

This section gives guidelines for the selection and design of a pilot study area as a first step to upgrading the network. A pilot study area can be used to:

- demonstrate the principle of zoning
- demonstrate the principle of night-flow monitoring and data collection
- act as a test site for methodologies and technologies
- demonstrate results
- collect data
- train staff
- assess performance indicators

The pilot study can be used to assess both real and apparent losses.

5.7.1   Selection

- Recommended size of area: 1000–3000 connections (10–30 km network).
- Representative of the network features (population profile, leakage, illegal connections).
- Hydraulically sound (suitable for zoning and closing some boundary valves).

- Good records of pipes and fittings (position, size, material etc.).
- Good customer records.

5.7.2   Implementation

Implement zone boundary:
- select site for zone input flow meter (and PRV if needed later)
- select size of meter based on estimate of population multiplied by per capita consumption
- install meter
- close zone boundary valves or install further input/output meters
- monitor pressures at critical points at peak demand while closing valves
- select pressure monitoring points for pressure survey
- select and install flow/pressure data logger on meter(s)
- record zone features (boundary position, meter positions, pressure monitoring points etc.) on plans
- calculate no. of connections and population
- identify trade/commercial customers

5.7.3   Studies

Real losses

Collect data for:

- zone total net flow (input flow minus output flow) in litres/hour or m³/day
- zone flow profile (24 hour flow pattern)
- zone night flow (e.g. period from 0200-0400) in litres/second
- average trade/commercial use (from billing records)
- customer night use (from sample of meters read at night)

Then estimate *leakage* from above information (in litres/hour or m³/hour)

Apparent losses

Meter error: check accuracy of a sample of customer meters, either by:

- removing meters temporarily and checking on a test rig or
- installing a 'master' check meter in line with customer meter to compare readings

Meter under-registration: consider installing a sample of 'C' or 'D' class meters and note effect on consumption/revenue.

Theft/illegal connections: conduct a survey of all households or a sample area to:

- assess proportion of stopped/damaged/bypassed meters
- assess number of illegal connections

Use this information to estimate actual night consumption and re-assess leakage figure

Customer education: formulate a customer awareness programme for demand reduction, water saving etc.. Visit a sample of customers to explain the concepts etc.

Levels of service (performance indicators):

1 Pressure
- monitor pressure at input/output meters and at critical points (high spots, critical customers, ends of line etc.)
- set pressure standard of service (e.g. 20m) and monitor frequency of failure to meet standard

2 Flow
- set flow standard of service and monitor frequency of failure to meet standard

3 Customer complaints
Record number of complaints – categorise as:
- water quality
- no water
- below standard of service for pressure/flow

Leak detection:

Test methods and equipment:

- leak localisers
- step test
- sounding, ground microphone, leak noise correlator

Find and repair leaks
Read meter(s) and calculate reduction in leakage volume
Note improvement to pressure

Pressure reduction:

Conduct pressure survey after leak detection and repair
Consider scope for pressure reduction in whole or part zone

Training and awareness:

Use pilot zone to train staff on principles of:

- flow and pressure monitoring
- data capture and analysis
- leak detection and location technology
- customer relations

Use results, data, and experience of zone studies to introduce regular 'staff awareness' workshops, where staff can exchange experiences and knowledge.

5.8   MAINS RENEWAL AND REHABILITATION TO REDUCE LEAKAGE

This section is based on the experience of practitioners in the UK water industry [1], which can be applied to all networks.

Most water supply organisations regularly carry out work to renew or rehabilitate their water distribution networks. If they did not, the pipe network would continue to age and deteriorate, resulting in increasingly higher maintenance costs to carry out repairs, in order to maintain levels of service to customers. However, age alone is not a reliable indicator of the need to replace a section of main. Often cast-iron mains over 100 years old can produce a better performance than spun-iron mains laid in the period from 1945 to 1970. Burst frequency is also a function of pipe material, diameter, ground conditions, quality of pipes and laying, and other external factors.

If a pipe network is, on average 50 years old, and the aim is to prevent it aging further, then at least 2% of the network will have to be rehabilitated each year. This will equate to a length, which has to be selected based on a number of condition and performance factors of which leakage will be one.

The primary justification for main renewal and rehabilitation is usually one of the following:

- The internal condition of the main is affecting the quality of the water delivered through it. This is often the case with corrosion of cast or ductile iron pipes, which have no internal protection.
- The internal bore of the main has reduced due to corrosion, or a build up of deposits, so that it is no longer capable of carrying sufficient flow
- The pipe wall has weakened and is no longer capable of withstanding the internal pressure of water, or it has insufficient beam strength to withstand traffic loading. This is often the case with asbestos cement pipes laid in aggressive ground.

- Some external factor has resulted in the main being unable to fulfil its current duty.

It is unusual for mains to be replaced solely on the grounds of leakage reduction. Customer levels of service and operating costs are the primary drivers. In all cases, however, the impact on leakage should be assessed as part of the justification process for each section of main.

The impact on leakage will depend entirely on the extent to which the old main was contributing to the overall leakage level, and then on the technique chosen to rehabilitate the network. If mains are replaced with new ones, then leakage on the main will be reduced significantly, although not eliminated altogether. However, unless the service connections are also renewed, there could actually be an adverse effect due to increased pressure (due to the extra carrying capacity) causing leaks on services to flow at a higher rate. Mains relining can also result in higher leakage due to the scraping process causing damage to pipe joints, service connections, and the pipe wall. If the main is slip lined with a new plastic pipe, then this will not be a problem, but if the main is coated with cement mortar or epoxy resin, then the rehabilitated main may leak at a higher rate than before.

In overall terms the experience of the author (Trow) shows that unless main rehabilitation and renewal is targeted specifically to reduce leakage then it will have a neutral effect on leakage levels. The benefits gained in some projects, will be offset by the increased leakage caused by others, and in any event the small percentage of mains which are rehabilitated each year are statistically unlikely to be the source of the leakage problem.

It is often assumed that mains which are in poor condition because they suffer from internal corrosion, or because they cause levels of service problems to customers, are also be the prime source of leakage. However, evidence from the UK suggests that this is not necessarily the case. Also, there is evidence to show that there is little or no correlation between burst frequency and background leakage. Areas with high burst frequency can have a low background leakage and vice versa. This may be due to the fact that high burst frequencies tend to occur in smaller diameter mains with low beam strength. Background leakage will tend to be more of a problem on larger diameter mains, and on service connections. Therefore, each section of main has to be assessed, and any policy based on generalisations is unlikely to be cost effective, and could actually produce little or no leakage reduction.

The cost of main rehabilitation in terms of its benefit, is likely to be in excess of £10m (€/US$16.5m) per Ml/d of leakage saved, unless the expenditure is targeted specifically at mains with the highest leakage rates. The cost of so-called 'blanket' renewal is almost certainly more than any other demand management or supply augmentation solution, and so it will not be a cost effective option.

5.8.1   Selecting mains for renewal

If mains rehabilitation is to be part of the leakage management strategy then it should be targeted. The aim is to identify those mains which make the major contribution to leakage levels in a supply zone, and then to find the most appropriate technique to renew them. This investigation has a cost, and therefore a balance has to be found between the expenditure on analysis and design, and on the actual cost of the main replacement. If insufficient effort is made on the preparatory stage, then mains will be replaced with little benefit, whereas too much investigation will add to the overall cost unnecessarily.

The following steps should be followed.

1. Identify those mains clearly in need of replacement

The first step is to examine records of mains failures and to consult with the local operations staff to identify those sections of main with a history of bursts and leaks, where repairs are carried out regularly. These can be prioritised simply in terms of bursts/km/year.

There will be a break-even point depending on the value of the water, the frequency of bursts, the volume lost from each burst, and the cost of continuing-with repairs. For example, the leakage from a burst on a 100 mm water main losing 5.5m^3/hr at a pressure of 50 m, with water costing £0.10 (€/US$0.16) /m^3, will have a net present value (over 20 years) of £56 000 (€/US$92 400). To eliminate this burst by mains replacement would cost £320 000 (€/US$534 600).

Figure 5.3 shows an example from the UK [1] of a break-even analysis for one particular water supplier:

2. Identify areas of high leakage

The next step is to identify those areas which have a high level of leakage, after leakage detection and repair work has been carried out. This is best done for each DMA where they have been established. The areas should be prioritized according to their infrastructure condition factor (ICF – see Chapter 4), or simply in terms of the background leakage in litres/property/day or m^3/km/day. Each area should then be investigated to determine the primary sources of leakage. Any leaks which can be detected by the usual location methods will have already been identified, so some form of step testing or sub-metering should be used. The aim is to measure the leakage

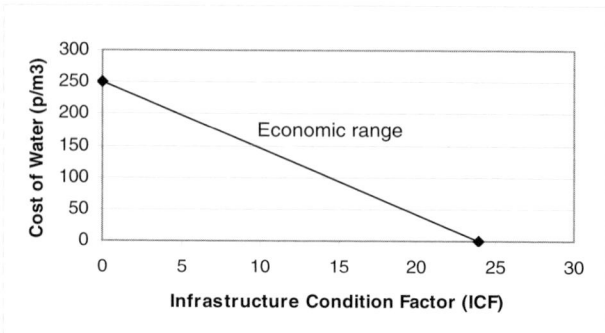

Figure 5.3 The economic range for selective mains replacements to reduce burst frequency background leakage as measured by ICF.

on sections of main. Ideally each street should be examined, but if this is not possible, then the leakage should be narrowed down to as small an area as possible.

3. Cost–benefit analysis

Within the DMA, each sub-area can be assessed to determine whether it is cost effective to replace the mains to remove the background leakage. The leakage rate will vary considerably from one section of main to another. In carrying out the investigation, a burst which was not located by other means may actually be found

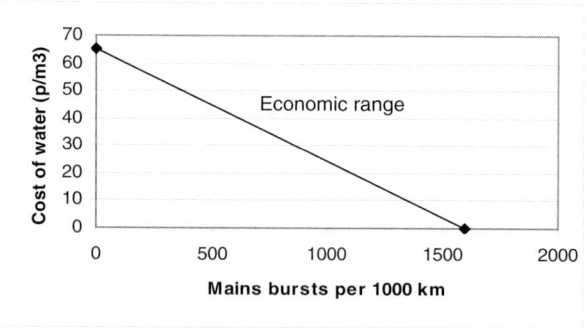

Figure 5.4 The economic range for selective mains replacements to reduce burst frequency.

and repaired, eliminating the need to replace the whole section of main. Figure 5.4 shows an example of an economic range in which it is cost effective to carry out the replacement works solely justified on leakage reduction.

4. Consider other benefits

When carrying out the cost-benefit analysis it may be possible to bring other benefits into the equation and to give them a value. For example, if the replacement is aimed to reduce bursts there will be a benefit from not having to carry on with repair costs, and there is a benefit in terms of customer service.

5. Design the scheme

A package should be produced including a plan and all other relevant data.

6. Project management

It is vitally important to ensure that the scheme realises the projected benefits by ensuring that:

- The old mains are completely removed from the distribution network. In some cases, the old main may be left in service because it is difficult to transfer some of the service connections, or because it cannot be cut and capped. However, unless it is completely disconnected, it will continue to leak. Worse still, if it is removed from the mains records plans or GIS system, then future leak location exercises will not include the section of old main.
- The quality of materials, and workmanship should be to a high standard

When main replacement is being carried out it is worth considering the following points:

- Unless service connections are also renewed then the benefit will be significantly less.
- Service connections should be installed using high quality materials which will be resistant to future leaks and bursts.
- Although the new mains may be capable of withstanding 10 to 16 bar of pressure, the pressure should still be managed to the minimum necessary to meet customer expectations.
- If the main cannot be pressure reduced, then it worth considering whether each individual service connection should be pressure reduced in conjunction with the work being carried out.

5.9   REFERENCES

1   Trow, SW (1997) 'Selecting water mains for renewal to reduce leakage levels'. Paper presented at CIWEM Northern Region Conference, Wakefield, October.

6

Leakage monitoring and control

6.1 CRITERIA AND CHOICE OF METHODOLOGY

The criteria used to develop the most appropriate strategy to address the components of water loss have been addressed in Chapter 4. This chapter deals with selecting the most appropriate methodology and technology to support a leakage management strategy.

Leakage management can be classified into two groups:

- passive (reactive) leakage control
- active leakage control (ALC)

6.1.1 Passive control

Passive control is reacting to reported bursts or a drop in pressure, usually reported by customers or noted by the company's own staff while carrying out duties other than leak detection. The method can be justified in areas with plentiful or low cost supplies. Often practised in less-developed supply systems where the occurrence of underground leakage is less well understood, it is the first step to improvement (i.e. to make sure all visible leaks are repaired). Except in exceptional circumstances leakage will continue to rise under passive control.

© 2003 IWA Publishing. *Losses in Water Distribution Networks* by Malcolm Farley and Stuart Trow. ISBN: 1 900222 11 6

6.1.2   Active leakage control

Active leakage control is when company staff are deployed to find leaks which
have not been reported by customers or other means
    The main methods of ALC are:

- regular survey
- leakage monitoring

Regular survey

Regular survey is a method of starting at one end of the distribution system and
proceeding to the other using one of the following techniques:

- listening for leaks on pipework and fittings
- reading metered flows into temporarily-zoned areas to identify high-volume
  night flows
- using clusters of noise loggers (leak localising)

Leakage monitoring

Leakage monitoring is flow monitoring into zones or districts to measure leakage
and to prioritise leak detection activities. This has now become one of the most
cost-effective activities (and the one most widely practised) for leakage management.

6.1.3   Selection of the most appropriate policy

The most appropriate leakage control policy will mainly be dictated by the
characteristics of the network and local conditions, which may include financial
constraints on equipment and other resources. Staffing resources are relevant, as a
labour intensive methodology may be suitable if manpower is plentiful and cheap.
If the geology of the area allows a high proportion of leaks to appear at the surface
(e.g. parts of the Middle East and Australia), a policy of regular survey followed by
rapid repair may be adequate. Where some leaks fail to appear at the surface,
however, a more intensive policy of leakage monitoring is required.
    The main factor governing choice, however, is the value of the water, which
determines whether a particular methodology is economic for the savings achieved.
A low activity method, such as repair of visible leaks only, may be cost-effective in
supply areas where water is plentiful and cheap to produce. On the other hand,
countries which have a high cost of production and supply, like the Gulf States, can
justify a much higher level of activity, like continual flow monitoring, or even
telemetry systems, to warn of a burst or leakage occurring.

In many developing countries the method of leakage control is usually passive, or low activity, mending only visible leaks or conducting regular surveys of the network with acoustic or electronic apparatus.

The volume lost from a leak is a combination of the flow rate together with the awareness time and the time taken for location and repair:

- *awareness time* – the average time from the start of a leak until the water company becomes aware of its existence
- *location time* – the average time taken to locate the position of the leak
- *repair time* – the average time for the company to shut off and repair the leak

The main effect of an ALC policy is reducing the average *duration* of leaks. The *awareness* time is influenced by the data capture method:

- telemetered flows – less than 1 day
- monthly night flow measurements – 14 days
- regular inspections – half the interval between inspections

The *location* time will be influenced by the nature and extent of monitoring systems, but mainly by the number of staff available and the equipment and technology at their disposal.

The *repair* time will normally be the same for leaks reported by customers or detected by proactive means. However, one of the main initiatives of a leakage management strategy should be to reduce the time taken to repair leaks once they have been located. Leak detection policies and procedures are detailed in 6.4.

6.2   SECTORISATION AND ZONAL MONITORING

A flow measuring system in a water distribution network should include not only measurement of total flows from source or treatment works (production), but also zone and district flows. This allows the engineer to understand and operate the system in smaller areas, and allows more precise demand prediction, leakage management and control to take place.

The measurement system must therefore be hierarchical, i.e. at a number of levels, beginning at production measurement, via zone and district measurement and ending at the consumer's meter or consumption estimate. This hierarchy is illustrated in Figure 6.1.

The system comprises:

- measurement of supply at the source or treatment works
- measurement of flow into supply zones, with geographic or hydraulic boundaries, usually 10 000–50 000 properties

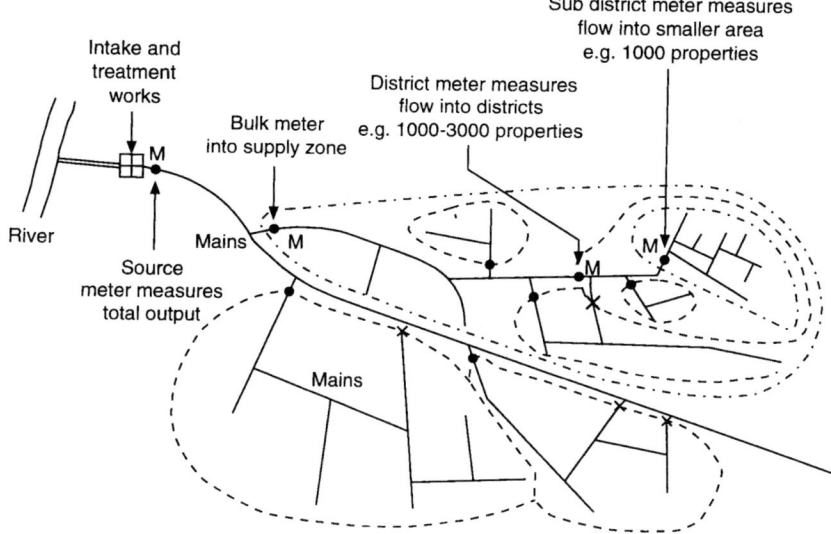

Figure 6.1 Metering hierarchy and DMA design options.

- flow monitoring into district meter areas (DMAs) of up to 3000 properties, with permanently closed boundary valves
- small leak location areas within each DMA, of around 500–1000 connections, where boundary valves remain open except during a leak location ('step test') exercise
- individual consumer meters, both domestic and commercial

6.2.1   Leakage monitoring

The technique of leakage monitoring is considered to be the major contributor to cost-effective and efficient leakage management. It is a methodology which can be applied to all networks. Even in systems with supply deficiencies leakage monitoring zones can be introduced gradually. One zone at a time is created and leaks detected and repaired, before moving on to create the next zone. This systematic approach gradually improves the hydraulic characteristics of the network and improves supply.

Leakage monitoring requires the installation of flow meters at strategic points throughout the distribution system, each meter recording flows into a discrete district which has a defined and permanent boundary. Such a district is called a district meter area and the concept of design and operation of DMAs has been detailed in two WRc reports: *District Metering – System Design and Installation* [1], and

District Metering – System Operation [2]. The DMA concept was reviewed in *Managing Leakage* Report J [3], in 1994, and was updated by UKWIR in 1999 with the report *A Manual of DMA Practice* [4]

The design of a leakage monitoring system has two aims:

a) To divide the distribution network into a number of zones or DMAs, each with a defined and permanent boundary, so that night flows into each district can be regularly monitored, enabling the presence of unreported bursts and leakage to be identified and located.

b) To manage pressure in each district or group of districts so that the network is operated at the optimum level of pressure.

It therefore follows that a leakage monitoring system will comprise a number of districts where flow is measured by permanently installed flowmeters. In some cases the flowmeter installation will incorporate a pressure reducing valve.

Depending on the characteristics of the network, a DMA may be:

• supplied via single or multiple feeds
• a discrete area (i.e. no flow into adjacent DMAs)
• an area which cascades into an adjacent DMA

Figure 6.2 shows an example of the configuration of several DMA types within a 'water into supply' (WIS) zone boundary, and the DMA recording system:

• a transmission main DMA (501D04)
• a discrete DMA off a transmission main branch connection (501D03)
• a cascading DMA (501D02/501D01)

The DMA meters are sometimes linked to a central control station via telemetry so that flow data are continuously recorded. Caution is needed if telemetry is to be considered, as the cost can quickly escalate and exceed the value of the water lost. Analysis of these data, particularly of flow rates during the night, determines whether consumption in any one DMA has progressively and consistently increased, indicating a burst or undetected leakage.

It is important to understand the composition of night flow, as this will be made up of customer use as well as losses from the distribution system.

Culprit areas, i.e. ones showing a greater volume of night flow per connection than the others, can then be inspected more thoroughly by carrying out a *leak localising* exercise. Examples of these are:

• step test – a technique which requires the progressive isolation of sections of pipe by closing line valves, beginning at the pipes farthest away from the meter and ending at the pipe nearest the meter. During the test the flow rate through

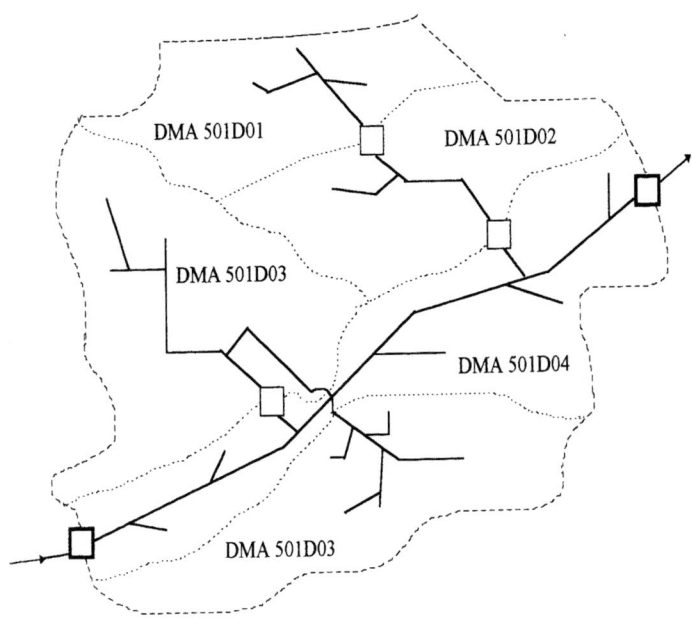

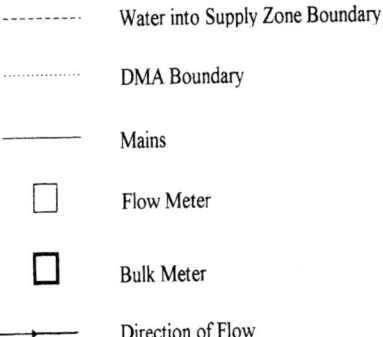

- - - - - - - -   Water into Supply Zone Boundary

. . . .   DMA Boundary

——————   Mains

☐   Flow Meter

■   Bulk Meter

——▸——   Direction of Flow

Figure 6.2 Configuration of several DMA types within a water into supply (WIS) zone.

the meter is observed and the times when each section of pipe is isolated is noted. A large decrease in flow, or "step", indicates a leak in the section of pipe which has just been isolated
- correlator survey
- acoustic logger survey

Inspectors can then be deployed to locate the precise leak position in the culprit section of pipe. Localising techniques are explained in more detail in section 6.3.

In developing countries it may be a difficult concept to introduce more closed valves in a system, particularly in low pressure areas or those with intermittent supplies. The benefits can be best demonstrated in pilot areas, when the benefits of leakage reduction on increased pressures and satisfied demand can be clearly seen by the customers. The benefits of pilot areas, and guidelines on their design, are described in section 5.7.

6.2.2   Monitoring in a DMA

Monitoring equipment

A DMA monitoring system involves:

- the acquisition of data
- the transmission of data from local sites to operations
- the archiving and presentation of data and reports

The system requires:

- a flow meter, for data gathering at source
- data logging equipment, for local site storage
- a communication system, for transferring data to the operator or by manual collection
- software, for archiving and presenting data

Figure 6.3 illustrates flow meter data being downloaded on-site from a data logger to a laptop computer

Flow meters

MECHANICAL METERS

Turbine meters, of the Helix (horizontal spindle) or Woltmann (vertical spindle) type, are the most widely used, and are readily available from several well-known manufacturers. This type of meter is also used almost exclusively in developing countries, as it is simple and cheap to manufacture locally, and easy to maintain. Its

Figure 6.3 Downloading flow data.

other main advantage is its wide flow range, allowing it to be used as a zone meter and for leakage detection. However, to protect the turbine from detritus in the pipeline, it should be installed with a strainer. To enhance data capture, the meter can be fitted with a pulse generator which produces pulses proportional to flow rate, allowing data to be recorded via a meter out-reader or data logger.

ELECTROMAGNETIC METERS

Although widely used for bulk flow measurement, electromagnetic meters are becoming increasingly popular for leakage management. The cost of these meters, which in the past has been prohibitive for mass installation, is reducing. They are very accurate, and, as they have no intrusive parts, are virtually maintenance free. Battery power supplies, particularly the advancement of solar powered cells, makes this type of installation increasingly worthwhile. They are also invaluable for situations such as a single-source DMA in a remote area.

ULTRASONIC METERS

Ultrasonic meters are rarely used for leakage monitoring, except in their portable form. Even in this form their cost prevents them from being a viable alternative to the insertion meter. However, ultrasonic meters can have a place in certain situations, such as remote sources, particularly when combined with battery power supply.

POINT VELOCITY METERS

Point velocity meters such as the insertion turbine and insertion electromagnetic meters, also have a place in network management. These are used at points in the system to measure flows where there is no existing meter, or to verify a meter already installed. They can be installed under pressure without interrupting supply, and can be transferred to other installation points as and when required. Insertion meters can be used for:

- measuring network flows for hydraulic modelling
- checking or calibrating existing meters
- permanent flowmeter installations in large or inaccessible mains
- flow monitoring in certain DMAs (e.g. large mains or difficult installations)

Chapter 5 (sections 5.2 and 5.3) details the meter types and installation requirements for DMA/zone meters, and Appendix B contains an example installation diagram and schedule of materials.

Data Capture Equipment

Data capture requires the use of pulse generators, pulse counters and data loggers (portable or with telemetry) to capture data from the DMA meter.

Pulse generators

These are solid state units attached to the register of a mechanical meter to provide a pulse output, in effect creating an electromechanical meter. All manufacturers of helix type meters supply units which are attached to the range of their helix meters.

Pulse counters/low cost data loggers

These units are usually low-cost single- or dual-channel data loggers used to record pulse output signals. Their main use is as a temporary installation data logger to record night flows over several nights as an indicator of leakage prior to a full data logging exercise. They are also suitable for household and non-household demand logging. Some loggers can be mounted directly onto helix meters, eliminating the need for a pulse generator.

Data loggers

Data loggers are used by most water companies to regularly monitor DMA flows on a 24-hour basis. They are dual- or multi-channel, designed for permanent installation if required.

An outline specification for this type of logger is as follows:

- logging interval from 1 second to 24 hours
- 380 days memory at 15 minute intervals
- submersible to IP68
- shock proof from 1.0 m drop
- temperature −10° C to + 60° C
- digital input to support:
  - range of mechanical helix meters
  - electromechanical insertion probes
  - EM insertion probes
  - contact closure pulse units
  - electronic pulse units used with a range of mechanical electronic pulse units used with a range of mechanical meters
- pressure input facility to support:
  - 0–25 bar for distribution monitoring sensors
  - 0–350 millibar for reservoir depth sensors
  - range of electrical inputs
- communications:
  - software selectable
  - local via RS232 interface, telemetry options
  - battery life 10 years (lithium chloride)
  - alarm capacity

It should be noted that some manufacturers supply data loggers with a sufficiently high specification to serve as both pulse counters and permanent data loggers. It is recommended that practitioners use the specifications above as a guide for selecting from a manufacturer the data logger suitable for the purpose.

All logger types should be compatible with the range of flowmeters which are being used by the international water industry for DMA monitoring. They should be able to receive digital and analogue signals so that they can also be used for pressure monitoring and level sensing, and with temporary insertion/clamp-on meter installations.

6.3   LEAK DETECTION AND LOCATION IN DMAS

There is a clear distinction between leak detection and leak location. Detection is the 'narrowing down' of a leak or leaks to a section of the pipe network. Leak detection activities may be carried out routinely, i.e. as a 'blanket' survey of the network, or in precise areas of the network, guided by the analysis of DMA data.

Leak location is the identification of the position of a leak prior to excavation and repair, although finding the exact location cannot be guaranteed. Location surveys can be carried out with or without prior leak detection activity.

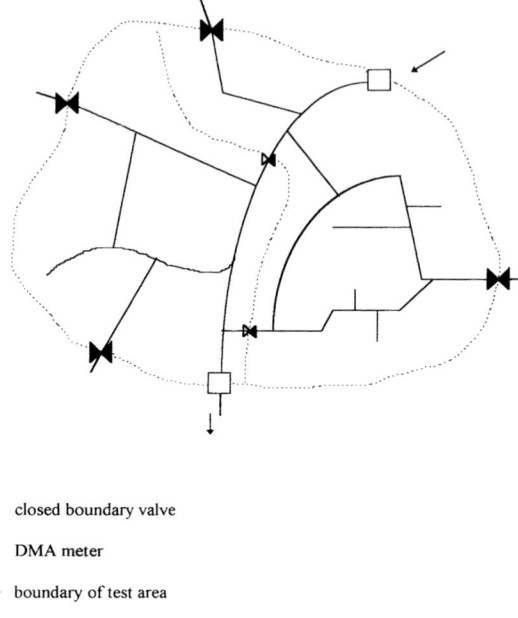

closed boundary valve

DMA meter

boundary of test area

temporarily closed valves

Figure 6.4 Subdividing a DMA by closing valves.

6.3.1   Leak detection techniques

There are a number of techniques to detect where leakage is taking place in the network. These include:

- sub-dividing DMAs into smaller areas by temporarily closing valves or by installing meters (see Figures. 6.4 and 6.5)
- variations of the traditional step-test
- the use of leak localisers (noise loggers)
- sounding surveys.

6.3.2   The choice between step testing and leak localising

The choice of detecting leaks by step testing or using leak localisers is decided by checking whether the network will support the step testing procedure. There may be reasons why a step test cannot be carried out, depending on the organisational structure and the network characteristics. It may:

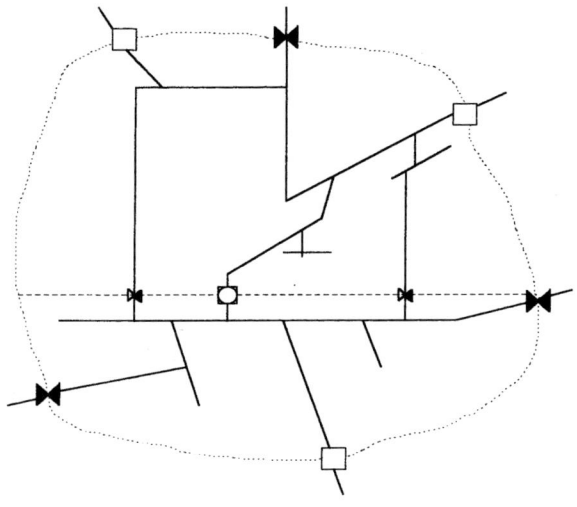

closed boundary valve

DMA meter

internal boundary meter

.... boundary of test area

temporarily closed valves

----- temporary internal boundary

Figure 6.5 Subdividing a DMA by metering.

- involve night work
- require two or more operators
- require customer warning
- cause water quality problems
- cause bursts in weak pipes
- cause inconvenience to night users

If these factors are a problem in a particular DMA, leak localisers should be used. However, step tests have advantages:

- results are immediate
- leak location can be done simultaneously

The choice of detection method will depend on:

- the configuration and characteristics of each DMA
- operator preferences
- water company policy

However there will clearly be situations where one of the two techniques is more appropriate or has particular advantages, and others where they can be used in tandem.

Step Testing

The techniques of step testing are well documented in *Leakage Control Policy and Practice* (Report 26) [5]. The procedures for setting up and operating step test areas are still valid today. However, the technology for both monitoring and recording flows has advanced since the report was published – step test areas are now generally smaller (500–1000 properties), and are usually incorporated within larger DMAs, using the DMA meter for monitoring flows into the DMA and for carrying out step tests in each area.

The procedure is summarised below, with the terminology updated.

1 Establish a step test area:
- determine the number of properties in the area
- determine the number of metered customers who use water at night
- estimate the number of un-metered non-domestic customers, taking note of those likely to use water at night
- check condition of valves to be operated during the test
- allocate numbers to valves and note if they are clockwise or anti-clockwise closing
2 Prepare a plan of the step test area, to show:
- road names and layout of pipes
- meter installations and valves
- boundary valves (closed to isolate the area from the DMA)
- circulating valves (closed to remove loops, to create a tree and branch network)
- step valves (operated during the step-test)
- all other valves, not used during the test, to avoid opening in error (e.g. DMA boundary valves)
- positions and details of commercial customers, with an estimate of their night use (to help later analysis of step test data)
- valve numbers, status (closed or open), and direction of closing

A typical sequence, practised by UK water companies, is as follows:

- step test at night, unless day-time is unavoidable
- close all circulating valves prior to test
- attach logger to meter, programmed for 1-minute intervals
- start at point furthest away from meter and close the step valve, recording the time of shut and valve details
- sound each valve for tightness
- work back towards meter
- each step is shut long enough to see the impact at the meter (typically 10 minutes)
- if urban area with bulk users (e.g. hospitals) back-feed the area as each section is shut
- some practitioners re-open step valves immediately after each test so that supplies are only lost for the duration of the valve operation
- re-open in reverse order, opening each valve slowly to avoid bursts
- sound leaky sections on the same night or the following day

A typical time requirement for a DMA with 25 steps is 4 hours 10 min. (25 × 10 min.) plus operating time. The use of *remote step testing* equipment speeds up the process – a leaky pipe section can be seen immediately by the team, and each step takes 2–3 minutes rather than 10 minutes (see section 6.5.1).

Leak localising (noise logging)

Although flow monitoring can localise a burst to a section of main, distances can be large, and some form of localising or pre-location technique is required. This technique has become popular over the last few years, and practitioners are increasingly using the technique as an alternative to step testing. The technique can also be used as a routine survey to 'sweep' a zone. The equipment is supplied by several manufacturers. The system consists of a set of six or more microphones, each incorporating a logger (see Figure 6.6). The units are magnetised to ensure contact between the sensor and the metal. Units are installed on a group of adjacent fittings (usually valves or hydrants), and are set to switch on automatically at a predetermined time. The loggers listen for and record the constant source of noise generated by a leak, usually over a two hour period. Because they are set out during daytime and record noise levels automatically at night, they have an advantage in busy or dangerous areas. Analysis of the readings is done by comparison of sound level and sound spread recorded at each logger. This indicates whether there is a consistent anomaly at one or more of the fittings, requiring closer inspection in the vicinity of that hydrant. Proximity to a leak is typically represented by a high decibel level and narrow noise spread. However, when analysing logger results they should

Figure 6.6 Installing a noise logger.

be compared with each other and not in isolation, to compare the significance of the results from the group. One localising system allows the localiser loggers to be 'surveyed' from a passing vehicle equipped with a control unit and receiver. Only those loggers which give an anomalous signal are investigated further. There is also now available a 'correlating logger', which records the leak sound into the logger memory, and which correlates and displays leak locations between loggers (see section 6.3.3).

Leak localising in transmission mains

A variation of the localising system has been developed to localise leaks in long lengths of transmission main. The sensor is attached to sounding points on the main which allow access to the water column. The sensor picks up leak noise from the water itself rather than the pipe material – it can therefore detect low frequency noise over long distances. Recommended spacing is at 500 m–1000 m intervals, although for metallic materials 2 km intervals are possible.

The cost of the leak localiser system is around £6000–7000 (€/US$9900–11 500) for a set of 6 sensor/logger units. No specialist skills are required and the only operating costs are staff time for setting out and collection, and downloading the data.

One obstacle to using this technique is the lack of fittings on most transmission mains for attaching noise loggers. Sounding points at 1 km intervals can be installed for this purpose – they are suitable for both noise logging and correlating.

Sounding survey

Sounding is the systematic survey of a DMA, listening for leak noises on valves, hydrants, stop-taps or at the ground surface above the line of the pipe. A sounding survey can be done either as the follow-up stage to a leak detection exercise, or as a blanket survey of the whole DMA.

Although blanket sounding can be inefficient in terms of focusing on leaky areas, it does provide a systematic examination of the DMA, such as when a DMA is first commissioned. It also allows other non-leak faults to be identified.

Sounding surveys are carried out using a variety of types of equipment. These are:

- basic listening stick
- electronic listening stick
- ground microphone
- leak noise correlator (survey mode)

The basic instrument is the *sounding stick* (see Figure 6.7) which is used either as a simple acoustic instrument, or electronically amplified. This technique is still widely preferred by the majority of practitioners, and is used for:

- blanket surveys, sounding on all fittings
- sounding on valves and hydrants

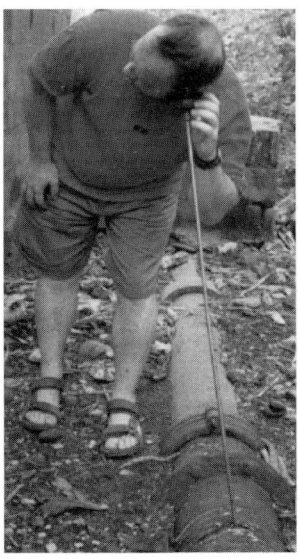

Figure 6.7 Sounding stick.

Figure 6.8 Ground microphone.

- confirming the position of a leak found by other instruments (ground microphone, leak noise correlator)

The *ground microphone* (see Figure 6.8) can be assembled for use in either of two modes, contact mode and survey mode. Contact mode is for sounding on fittings, similar to an electronic listening stick. Survey mode is used to search for leaks on lengths of pipeline between fittings. The technique involves placing the microphone on the ground at intervals along the line of the pipe and noting changes in sound amplification as the microphone nears the leak position. When a leak is detected the ground microphone is used in either mode for leak location.

The *leak noise correlator* (see Figure 6.9) is the most sophisticated of the acoustic leak location instruments. Instead of depending on the noise level of the leak for its location, it relies on the velocity of sound made by the leak as it travels along the pipe wall towards each of two microphones placed on conveniently spaced fittings (maximum 500 m apart). Hydrophones can also be used to enhance the leak sound in plastic pipes or large pipes. There is no doubt that the latest versions of the correlator can accurately locate a leak (to within 1 m) in most sizes of pipe. The instrument is portable so can be operated by one man, and it has the capability for frequency selection and filtering. However, there are some leaks which the correlator has difficulty in locating, notably pipes which are low pressure, large diameter, non-metallic, and with infrequent contact points for microphone placement.

Figure 6.9 Using a leak noise correlator.

The correlator can be used in two modes:

- as a survey tool to detect leaks in sections of pipeline
- as a location tool to identify the leak position

When using the correlator as a survey tool a correlation peak shows that a leak noise is present. Before carrying out a correlator survey, a plan of the DMA should be prepared, showing the location of all valves and hydrants to be used in the survey. These should be numbered on the plan, and a table prepared showing lengths of pipe by end numbers, and the estimated distance between locations. An example of a two-man correlator survey is as follows.

One operator carries the correlator and microphone and a copy of the plan. The other operator carries the second microphone. Both operators carry radios or mobile phones. The sequence is:

- Operator 1 attaches correlator and microphone to first fitting
- Operator 2 attaches microphone to second fitting and tells operator 1
- Operator 1 carries out a correlation run and any resulting leak(s) noted in the table
- Operator 1 radios Operator 2 to say when to move on to the next fitting
- Operator 1 repositions the correlator for the next run
- The sequence is repeated until the area is complete

The correlator is used in location mode on those sections which appear to have leaks. The advantages of the correlator survey are:

- it is unaffected by ambient noise
- unlike step testing, it can be done during the day
- it takes only a little more time than valve and hydrant sounding

Table 6.1 lists suggested procedures for different types of DMA, using sounding, step testing, leak localising, and leak noise correlator (LNC).

6.3.3 Leak location techniques

Leak location is the pinpointing of a leak position once it has been identified from one of the detection techniques. Equipment for leak location varies from the simple sounding stick, or stethoscope, to the highly sophisticated leak noise correlator. In between these extremes lies a range of equipment suitable for all systems. As well as the acoustic sounding stick, an electronically amplified version is available – however both types depend on the trained ear of the inspector.

Less dependent on the human factor is the ground microphone, which comprises a microphone and a recorder giving a visual display of the noise volume. The

Table 6.1 Techniques used by DMA type.

DMA Type	First pass investigation	Second pass investigation	Follow up	Comments
Town centre	Correlator/ acoustic loggers	Sounding/ acoustic loggers	Correlator survey	Investigations normally carried out at night
Large urban	Acoustic loggers or correlator survey	Sounding/leak noise correlation to pinpoint leaks	Correlator survey	
Small urban	Sounding	Correlator/sounding	Correlator survey/stick sounding	
Large rural	Step-test/ acoustic loggers	Localisers/correlator	Check night flow	
Small rural	Sounding	Correlator/sounding	Correlator survey/sounding	

microphone is placed on a pipe or fitting, or on the ground surface above the main, and the network is progressively surveyed for leak noise.

Whichever piece of equipment is chosen, all are used in exactly the same way as for leak detection, except that they are used more intensively in smaller areas of the network to track down the leak position. In the case of the correlator, 'location' mode is used instead of 'survey' mode.

A further enhancement to leak location by correlation is the combination of the correlation technique with acoustic loggers. This is a set of acoustic loggers programmed to record the leak sound into the logger memory. Loggers are deployed during normal working hours. When retrieved the recorded sound is automatically correlated to display the locations of leaks between all combinations of loggers.

There are also a number of other acoustic and non-acoustic location methods, which are usually used when acoustic methods fail to find the leak. Besides the well-established gas tracing technology, several new technologies are being developed and trialled by the water industry.

Gas injection

Gas injection and tracing techniques are used less frequently for leak location, mainly because the other techniques are successful in most cases. For the difficult leaks, however, particularly those in low pressure, non-metallic transmission mains, gas injection is the next choice. An inert gas is injected into the pipeline, and is traced as it comes out of solution at the leak point. Because of the equipment needed to inject and trace the gas, however, it is more cumbersome than correlation, and is usually carried out by a specialist contractor. The most common tracer gases used are *sulphur hexafluoride* (SF_6), industrial *hydrogen* (95% nitrogen, 5% hydrogen) and *helium.* The main disadvantage of the SF_6 technique is that bar-holes have to be made in the ground at 1 m intervals along the line of the pipe, to allow the gas to collect and be traced.

Figure 6.10 The hydrogen gas technique.

The main advantage of the hydrogen or helium technique is the speed of tracing. The gas diffuses through the soil (and asphalt and concrete, but less quickly) as it comes out of solution, and rises to the surface, eliminating the need for bar-holing. From practical experience the main applications of the *hydrogen* technique are:

- finding multiple small background leaks in a single section of pipe, e.g. during a step test
- finding leaks in service pipes, which are relatively close to the surface, and which can contain unexpected loops and bends making accurate correlation difficult

The tracer gas is injected into the network via a hydrant upstream of the suspected leak. The operator walks along the line of the pipe with a sensor, which is either hand held or a surface probe. The sensor is microelectronic, and the manufacturer claims that no maintenance is required. The hand-held sensor is used for tracing on accessible pipes or fittings – the surface probe for tracing the gas close to the ground, along the line of the pipe. Figure 6.10 shows the technique being used in both modes.

The *helium* technique requires less gas, and it does not have to be mixed with nitrogen to make it safe. Like hydrogen, it is suitable for service pipes but is not recommended for mains due to the large quantity of gas needed.

Figure 6.11 In-pipe acoustic technology developed by Tokyo Water Works.

Other techniques

Several innovative techniques have been tried by the water industry [6] as an alternative to the conventional techniques described above, usually to find difficult leaks, especially in transmission mains. The techniques include ground-penetrating radar (GPR), thermal imaging, and in-pipe acoustic technology. This latter technique requires under- pressure insertion of a microphone, or pair of microphones (see Figure 6.11) into the main. The velocity of water along the main, sometimes assisted by a drogue, carries the microphone towards the leak position. As the microphone passes the leak position the noise is and the position is recorded. This technique is now fully developed and is commercially available. Some of the other alternative techniques are fully developed, others are still being trialled in the industry. Several water companies claim success with ground radar, and with thermal imaging, using an aircraft mounted camera to over-fly rural transmission mains.

6.3.4   Leak detection in transmission (trunk) mains

These larger mains can pose specific problems for leak location. Transmission mains are defined as:

• mains which transfer water between sources, treatment works, and service reservoirs

- mains which transfer water from service reservoirs to the distribution network
- distribution mains which are greater than 300 mm, and outside the boundary of a DMA. (Some companies incorporate large mains in this category, particularly those with multiple branch connections, into the DMA system. Other companies treat large mains, usually those with few branch connections, as a 'transmission main DMA', with meters positioned where the main crosses the DMA boundary and at branch connections)

While transmission main bursts are usually visible and dealt with quickly, leaks from joints on transmission mains can sometimes go undetected for years, leading to huge volume losses. However such leaks, particularly in rural transmission mains, are notoriously difficult to detect and locate. This is because transmission mains usually have one or more of the following characteristics:

- low-pressure
- low noise frequency
- large-diameter
- non-metallic materials
- infrequent contact points for acoustic location

The worst scenario is, of course, a combination of all these characteristics, which is not uncommon. Rural transmission mains have additional disadvantages:

- they are often remote
- they are laid over long distances
- they can lie in difficult terrain
- their exact position is sometimes uncertain

A survey of traditional and new technology for detecting and locating leaks in transmission mains, carried out for the UK water industry in 1999 concluded that:

1 The development of new monitoring and location technology for transmission main leakage should be viewed against a background of state of the art flow measurement and correlator technology. Particular note should be made of claims of 'advanced' correlator technology by manufacturers new to the market.
2 Monitoring flows to detect leakage in transmission mains is restricted to measurement of velocity differences between pairs of portable meters, insertion EM or clamp-on ultrasonic.
3 The most common alternative technologies currently used by UK water industry practitioners to support the correlator are:
   - leak localising
   - ground-penetrating radar (GPR)
   - hydrogen gas tracing

4 A suitable combination for a system of leak detection (or localising) and location is:
   - detection – GPR or leak localisers
   - location – correlator or in-pipe technology
5 Thermal imaging using fixed wing aircraft or helicopter is seen as an expensive but fast technology for scanning rural transmission mains – it may be cost effective for extreme or isolated situations.
6 New technologies being trialled and developed further are:
   - radar technology to detect leak vibrations
   - hydrophones, using principles of sonar and acoustic behaviour of sound in water developed to detect submarines
   - satellite imaging, which is awaiting availability of appropriate resolution images at realistic prices

6.4   LEAK DETECTION POLICIES AND PROCEDURES

The previous sections in this chapter refer to the techniques and technology for leakage management. This section gives guidance on the policy issues and human resource issues which affect active leakage control operations. Active leakage control consists of two distinct, but related, activities:

- leak location – the work involved in finding leaks in an area which has excess leakage over and above a set target level
- leak repair – the work required to effect a satisfactory repair within a reasonable time

Some general issues apply to both activities. In areas where water is cheap and plentiful an appropriate strategy may be to only repair those reported leaks which are brought to the attention of the water supplier, and not to go out actively looking for unreported leaks. In this case the only issue for leak location is the technique to be used to pinpoint the leak before the hole is excavated to carry out the repair. The level of leakage repair effort will tend to be similar from one year to another, although there may be seasonal differences, and so there are fewer issues with leak repair policies. This section deals with the policies and procedures associated with an active leakage control strategy.

6.4.1   Staffing policy

The first consideration for the water supplier before embarking on a major leakage reduction programme is whether or not the number of staff with the relevant technical and managerial experience to undertake the work are currently employed. In a well-run organisation it is unlikely that all necessary resources will be available 'in house'. Therefore, there is a key decision to be made. Should staff resources be supplemented by recruiting additional employees, or should all or part of the work be carried out by external contractors? This decision applies to both the leak location element and the repair work, and may also apply to the general management and technical issues.

Even if the decision is taken to contract out the implementation of the project in its entirety, there will still be an additional workload associated with contract management, coordinating the impact on other distribution system operations and works, and providing the contractor with local knowledge of the distribution system and the customers supplied.

At the other extreme, if the decision is taken to keep as much 'in-house' as possible, there is unlikely to be the correct balance of management, operational and technical staff, and there also has to be a balance between detection and repair effort. There is also likely to be a seasonal effect tending to drive up leakage at certain times of year. Extra resource may be needed to cover short term peaks in workload. Therefore, it is most probable that some additional external staff will be needed.

If the water supplier decides to take on additional staff, rather than use contractors, there are a number of human resource (HR) issues to consider. What happens to the extra staff at the end of the programme? Some additional staff will be required to maintain leakage at the lower level. Some will be allowed to leave the organisation through retirement. It is useful to undertake an age and capability profiling exercise before the leakage reduction programme to determine the likely trends in staff availability allowing for natural wastage. Taken with the estimated numbers of staff year on year, a recruitment and training plan can be devised.

The leakage reduction programme may have such an impact that it has to be considered as part of an overall organisational review which looks at all aspects of staff numbers, terms and conditions, management structure and staff incentives.

- New recruits will require training, if only induction into the policies and procedures of the particular company. However, it is also likely that technical skills training will be required. This will delay the commencement of the project and lead to additional strains on the company's management
- New staff will have to be provided with vehicles and equipment, which has to be procured and controlled, taking up more management time.

So, the issue seems to be not whether to contract out some of the work, but which elements of the work and how much. This balance will depend on the nature and extent of the leakage reduction programme, the profile of the staff available within the water supply organisation at the time, the availability and cost of external contractors, and the level of risk which the management is prepared to take.

There are also some issues associated with the urgency of the project. Keeping as much of the work as possible in-house may limit the impetus of the project. Using external contractors will provide an added stimulus to get the work done, particularly if there is some form of performance payment mechanism in the contract.

However, putting as much as possible out to contract may lead to friction between in-house staff working on the water distribution systems and the contractors. At the end of the project there may be a reluctance to accept the work that has been done by contract staff. This can result in facilities and zone boundaries not being maintained, or IT systems not being properly used to direct ongoing leak detection effort.

There are several key issues to be addressed in setting an appropriate policy for leak detection. These tend to fall into three main categories:

- the techniques to be employed
- the degree of effort required
- organisational issues:
  - the type of contract
  - management of leak detection operations

6.5   CHOICE OF LEAK DETECTION TECHNIQUE

The available techniques are set out in detail in previous sections. It is likely that a mix of these techniques will be employed and it is unwise to be too prescriptive about particular methods. An approach which seems to work best is one where the leak detection operative chooses the method to suit the nature of the district in which he/she is working, or the particular type of main. The following sections deals with the staffing issues of each technique.

6.5.1   Leak location in DMAs

The general approach will depend on the level of leakage above the target level. If it is such that it can be attributed to a large number of bursts and leaks, then it may be appropriate to survey the whole of the district. This may involve using the leak noise correlator to inspect every section of main, and sounding on every accessible stop tap.

A useful technique is to convert the level of excess leakage into the number of equivalent service pipe bursts (ESPBs), equivalent taps (ETs – 600 l/hr at 50m pressure), or equivalent mains bursts (EMBs). This is done by dividing the excess leakage (current leakage minus the target leakage – see section 6.5.4) by the volume of water lost through an average mains or service pipe burst. The average size of burst can be found by determining the reduction in leakage after repairs have been carried out, and is a key element of the economic level of leakage calculations. The number of ESPBs or EMBs will give an indication of the number of repairs which have to be carried out to meet the leakage target in that district. If it is a large number, the general survey approach is applicable. However, if there are only a few (say 1 to 5 ESPBs) then it is more efficient to narrow down the area in which the leak is running before attempting to pinpoint it.

After a general survey of a district, the next step usually involves finding the remaining leaks. The choice of technique will depend on how many leaks remain, and what other techniques have already been tried.

District metering in itself does not reduce leakage. Its purpose is to identify where leakage is occurring, and by doing so to make leak location more efficient. In areas which have district metering, regular analysis of the meter data allows those districts with excess leakage to be identified, and these can then be prioritised for inspection according to a number of factors e.g.:

- The level of the excess. A useful measure is the number of ESPBs per thousand properties in the district. This indicates the density of leaks in an area, and so the relative ease with which they will be found. Alternatively excess leakage may be reported in terms of litres/property/day for urban areas (ESPB/km for rural areas).
- The time since the last inspection. Using excess leakage alone will mean that some districts are inspected more often than others. Some which have a high rate of rise of leakage, will be inspected every few weeks, whereas others may not be scheduled for inspection for a year or two. There is merit in setting a maximum time between inspections to ensure that leaks in low priority areas do not go on indefinitely.
- Some economic measure taking into account the cost or value of water in the area. One simple method is to multiply the excess leakage by the unit cost of water in the area. This will give a product which places a value on the water which may be saved by leakage reduction.

This procedure of homing in on a leak uses the step testing and acoustic logging techniques:

Step testing

There are several forms which may be employed (see also section 6.3.2):

Transferring a small area or individual length of main from one district to another by opening and closing valves to alter the position of the district boundaries. If the flow rate between the district changes by more than would be expected for customer use alone, it will be because a leak is running in the area which has been transferred.

Individual sections of main may be isolated for short periods of time, depending on the statutory requirements for continuity of supply. Step test equipment can be used to assess the effect on the flow through the district meter while the valves are shut. A unit at the meter sends a signal to the operatives changing the valves. This allows the leak to be identified very quickly; in a matter of a few minutes, so that customer supplies are not affected greatly.

The other method is to split the district into sections, perhaps half, then quarter, and to determine which section the leak is in. This procedure can be followed until the culprit area is found, and then the individual streets can be isolated as set out above, The advantage is that the procedure is more efficient than shutting off every street, but it does have more of an impact on customer supplies.

Acoustic logging

Acoustic loggers can be deployed in an area to listen for leaks. They can either be used as survey tools, or they can be left in place permanently or semi-permanently. They indicate the vicinity of a leak and then other techniques can be used to pinpoint it.

6.5.2   Service reservoir monitoring

Service reservoirs should be monitored either by regular drop tests to assess water tightness or by continual monitoring of inflows outflows and stocks. Leakage can result from cracks and other defects in the walls or base of the reservoirs, or from defects on inlet and outlet mains. Other losses can occur due to overflows, particularly when the reservoir is not monitored by telemetry (or by more basic mechanisms such as the movement of a matchbox placed on the overflow weir or straw placed in the overflow pipe).

6.5.3   Transmission main leakage

Although it depends on the condition of the pipeline and the value of the water it carries, leak detection on transmission mains is usually expensive for the benefit which results in comparison to leak detection and repair on other elements of the network. It is difficult to detect leaks on transmission mains (see section 6.3.4). Even where meters are placed at the inlet and outlet of treatment works and service reservoirs, and each branch connection is metered, the meters are never 100% accurate. There will always be doubt as to whether the loss is a real losses, or is an apparent loss due to small differences in meter readings. Where there is a large difference, then this is usually due to one of the following reasons:

- *Meter error*: meters should be calibrated and cross checked with some other permanent measurement or by the use of temporary insertion meters or external ultrasonic meters.
- *Configuration*: the system characteristics may be different to those shown on the record plans.
- *Systematic error*: meter readings may be altered incorrectly in the process of the signal being converted into a flow rate or in the conversion of a value into other units of measurement. It is worth tracking key meter readings back to the actual ampere signal from the flow meter to verify if it corresponds to the value used in the calculations.
- *Unknown use*: customers may be taking water from the main without the knowledge of the water supplier. These may be illegal connections or they could be genuine supplies which have been overlooked in the water balance calculation.
- *Operational use*: sometimes, a valve may be left open on a section of transmission main to maintain a minimum flow rate in order to prevent water quality problems or to draw chlorinated water through the main.

The recommendation is to eliminate other causes of the apparent losses, before embarking on a programme of leak location. The following outline procedure is recommended for transmission mains:

- *Understand*: carry out site surveys to improve the understanding of the actual method of operation of the transmission main system, the nature of the offtakes, and how they are metered
- *Quantify*: carry out verification studies to calibrate meters, and to provide accurate estimates of any unmetered quantities
- *Report and plan*: analyse the data to determine possible causes of discrepancies and plan a programme of further action to find the cause of these anomalies.

If there is reasonable evidence to indicate the existence of a leak on a section of transmission main then there are several methods available including:

- leak noise correlators
- in-pipe acoustic technology
- walking and sounding
- ground probing radar
- aerial photography
- step testing
- pressure testing

6.5.4   Degree of effort

There are two fundamental issues:

How often should areas be revisited?

The time period between inspection of a particular district is known as the intervention interval. The average intervention interval is a key element of the overall strategy. It should be set using economic considerations as set out in Chapter 4. It will have a major impact on the level of resources required. The more regularly districts are re-visited the greater the number of staff required. The level of resource provided will also determine whether leakage is maintained at a given level or if it will be reduced, and the rate of reduction which can be achieved.

For any given level of resource, leak detection work should be scheduled according to some form of prioritisation as set out above.

For each district, an assessment is made of the cost of survey to bring leakage down to its set exit level. This may be based on previous experience, or some method related to the size and nature of the area. After each survey, the volume of water in excess of the leakage level is monitored each week. Once the cumulative value of the excess leakage (the average rate in m³/day times the unit monetary value) exceeds the survey cost, then another survey is carried out. The time between surveys will vary according to the rate at which the leakage rises and the value placed on the water at the particular time.

How much effort should be expended in each area before moving on to the next area?

Many aspects of leakage management follow laws of diminishing returns. This is also the case here. Once the initial leaks and bursts have been found, from what is often known as the first pass survey, then the more time which is spent in a district, the lower will be the return in terms of the number and quality of bursts which are found. So, the question is: at what stage should leak detection staff be moved to another area?

One method is to determine an 'exit' level for each district following a period of leak detection effort. As a general guide, based on UK experience, a two-person leakage detection team can survey a district of 2000 properties (with individual supply pipes and a frequency of connections to mains of say one every 15 to 20 metres) in 5 to 10 working days. For the first pass, they may be left for say 10 to 15 days in order to ensure that all locatable leaks have been found. The level of leakage which ensues following these repairs can then be set as the future exit level, or there may be a margin of say 10% above this level.

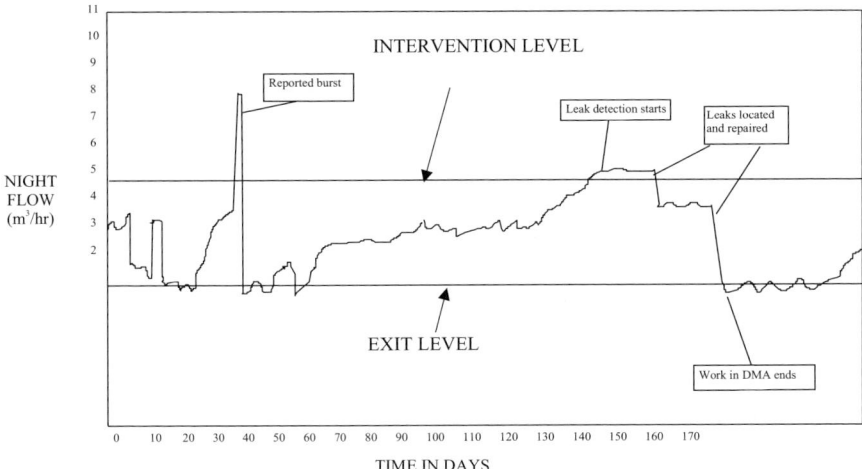

Figure 6.12 Example of intervention and exit levels.

Figure 6.12 shows how the leakage in a DMA varies between the intervention level and the exit level as new bursts occur, and after leakage detection and repair works are carried out.

Leakage in the district is then monitored and when it rises above the set 'entry' level or when it comes into the top priority districts or when the time since the previous intervention exceeds the set duration, then a second visit will be made to bring the leakage back to the exit level.

The value of this exit level in terms of litres/property/day will vary from one district to another. A useful technique is to compare districts in terms of their ICF (as set out in Chapter 4, section 4.7.6). Those districts with ICFs of between a half and twice the average may be seen as reasonable. However, it is possible that in those districts with ICFs over 2 that not all the locatable leakage has been found. In these cases, it may be worth re-visiting the district and using different techniques to locate leakage. Intensive forms of step testing may be used to identify the leakage level on each section of main. When this technique is used it sometimes finds leaks which would otherwise have gone unnoticed. Examples from the UK include:

- a leak on a main at the point where it crossed under a stream
- a leak on a main which ran through a sewer chamber on a busy traffic roundabout
- a split plastic pipe which only leaked at night when the pressure began to rise

There is also the possibility that leakage is occurring on apparatus which is not shown on record plans. Record plans are never 100% accurate, particularly when considering old mains which were thought to have been abandoned many years ago when the property they used to supply was demolished. In some cases, the connection to the existing network may still be in place.

In other cases, mains which have been replaced can still be connected. This is often found when it is difficult to transfer the service connections from the old main to the new one, and so some properties remain connected to the old main. Other examples are links which have been laid as temporary supplies while other work is undertaken on the network, but which are not removed on completion. *A Manual of DMA Practice* [4] lists methods of responding to DMA data and prioritising DMAs for leak detection.

6.5.5   Alternative policies without DMAs

If the area in question is not covered by district metering, it may still be possible to apply the same principles at a supply zone level. The larger area will mean it is difficult to identify individual leaks, and it will make leak location more difficult. However, a balance has to be struck between the investment cost of the district metering, and the ongoing cost of meter reading, and the savings which result from a more efficient operation. It is unlikely that 100% coverage of district metering can be justified economically. Indeed it may be technically impractical to adopt such a policy. Therefore, appropriate policies will have to be determined for those areas not included in DMAs.

These areas are subject to what is known as a policy of 'regular sounding', or regular survey. The frequency of the survey can be determined on economic principles and the strategy will include the number of staff required to maintain this inspection level. As it is difficult to estimate leakage in these areas they tend to be surveyed in rotation rather than on some form of prioritisation.

Economic frequencies of regular survey generally vary from a few weeks to about two years.

It may be possible to estimate leakage without DMAs, by a process of deducting the DMA flows from the total in the supply zone. In other cases 'dummy' DMAs are established, which are not metered but which are defined areas with known statistics so they are treated as DMAs in terms of leak detection activity.

6.6   TYPES OF CONTRACT

There are several benefits from using external contractors to carry out leak detection work:

- the ability to meet peaks and troughs in workload with fully experienced staff
- the ability to place staff where they are needed without having to consider issues of location and travel costs

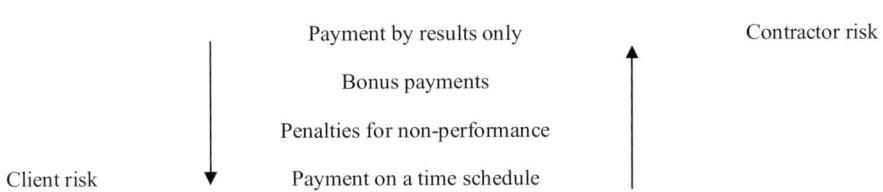

Figure 6.13 Different payment methods.

- the ability to provide an accurate estimate of the costs, and the activity rate or performance to be obtained
- the ability to focus staff on leakage detection without them being distracted by other operational duties
- the contractor can often bring new ideas gained from previous contracts with other water companies

There are many different types of contract, with different contract conditions [7]. The aim of this section is to give some guidance on the advantages and disadvantages of different payment methods. These can then be incorporated into standard contract terms which the water supplier may have used for other works contracts.

The payment methods vary from one in which the contractor is paid on a schedule of time rates, regardless of the achievements made, to one in which payment is only according to the volume of leakage saved in an area. Payment by results puts most of the risk on the contractor, whereas a time schedule puts the onus on the client. There are different forms of contract which aim to compromise between these two extremes (see Figure 6.13).

6.6.1   Payment by results

This method aims to relate payments to the contractor to the actual reduction in leakage [8]. The client is often more comfortable with this approach, as it is perceived that goals can be achieved at lower cost, because there is no 'wasted' expenditure. However, there are some drawbacks:

- The management cost tends to be much higher than that for a time-related payment method.
- The client must ensure that the methods of assessing leakage reduction are fair and reliable.
- The reduction in leakage can only be measured accurately after the repairs have been carried out. If there is a third party responsible for the repairs then this can be the cause of disputes, and delays in payments being received. Also, there will

be a tendency for new leaks to occur in the time between the detection contractor completing the work, and all the repairs being carried out.

- There is a tendency for leakage to rise again in a district after the find and fix work is completed. So, payments made for savings in leakage can be quickly wiped out.
- Care has to be taken to ensure that the apparent reduction in leakage is not due to some change in boundary valves, or a reduction in pressure, but that it is truly down to leaks actually located and repaired.
- The payment system may encourage contractors to count properties, meter commercial users, and carry out activities other than finding actual leaks.

To overcome some of these difficulties, performance contracts should consider the following:

- The size of the area which is monitored. It may be better to base payments on the leakage reduction achieved in the whole of a supply zone, even if it is completely covered by district metering. This ensures that there are no complications between leakage figures in adjacent districts, and that the demand into the whole zone can be monitored over a period of time.
- It may be appropriate to let this type of contract to a consortium comprising a leak detection contractor and an independent repair contractor.
- A payment should be made to the contractor for any better quality information which is obtained during the contract. These data may lead to lower estimates of leakage. 'Paper' savings, as they are sometimes called, can be often as valuable as actual savings in water. Such savings may come from better estimates of the number of properties supplied, information on the configuration of the network, data on customer use, and data on pressure. Some contracts include for a survey period, in which the contractor is paid to verify the data held by the water company before attempting to reduce leakage on a payment by results basis. The contractor is reimbursed for this work on a time schedule or by a fixed fee according to the size of the area.
- To avoid the delays while repairs are carried out, some contracts include a payment schedule according to the number and type of leak found. A points system may apply, with a certain number of points for a mains burst, fewer for a service burst and a minimum point score for a fitting leak. The contractor is expected to achieve a certain points value per team per week, and is paid according to the number of points scored. This is a useful form of contract where no district metering is in place.

In some cases, the payment mechanism is determined by the client, who then seeks competitive tenders from contracts. These may include prices for sample districts which are used to compare tenders. In other cases, the contractor is asked to submit a view on the payment mechanism as well as the rates.

For large contracts, some form of target cost approach may be used. The client provides data on current leakage rates, and the target to be achieved. The contractor sets out a price for carrying out all necessary work to achieve the target, and gives a target cost. Sometimes, the target cost, includes a provision for the value of the leaking water so that the time element between different tenders can be assessed. A longer contract period will result in more water being lost before the target is achieved.

If the actual cost is less than the target the client and contractor share the savings in an agreed way. However, if the target cost is exceeded, there are penalties which apply.

6.6.2   A hybrid system

To provide a shared risk between client and contractor, a hybrid system may be developed. This could include a fixed payment to the contractor for their attendance in operator hours, or fee for a survey of an area. The payment will relate to the number of teams employed, but it should be aimed to their cover costs only and not allow the contractor to make a profit. The profit has to be earned by achieving performance targets related to actual leakage reduction or to a points scoring system.

6.6.3   Schedule of rates contracts

These are usually simpler than payment by results contracts. The rates schedule will usually include a rate per day and/or per week for a single person working alone, or for a two-person team. It will also include for call out and standby charges, overtime rates, and equipment hire. There may be different categories of staff, e.g. a team leader rate, and an assistant rate.

The schedule of rates may also include a schedule of fixed fee rates for certain standard activities, in which case a system has to be established for recording and agreeing on activities carried out.

Some clients specify the number of staff required, and the only variable between contractors is the day rate. Other contracts ask the contractor to also estimate the number of staff required to achieve the target in a zone and to provide data on the nature of the zone and the current level of leakage.

Payment by results contracts should include a schedule of day rates to be used when the contractor is expected to carry out work in an area where the data normally used to assess the contractor's performance payment is unreliable.

6.7   MANAGEMENT OF LEAKAGE DETECTION OPERATIONS

This section describes the management techniques which should be applied to staff engaged on leak detection operations.

6.7.1   Staff management

Leak detection staff are in a position of trust in the organisation, whether they are employed directly or by a contractor. They spend most of their day alone away from the water company offices and depots. Their productivity is dependent not only on their own efforts, but also on the nature of the environment in which they are working and on the number of leaks which can be located. Staff must be trustworthy, as it is relatively easy for uncommitted staff to avoid working contracted hours or meeting minimum activity rates. Management techniques and controls will help, but there is no substitute for a dedicated and committed member of staff who enjoys the work of leak detection.

Time keeping

It is recommended that staff report in to a depot or office at the start of a shift and report back there at the end of it. Once staff have become experienced and they have the confidence of their line manager then it may be possible to drop one of these visits, so that they only call in once per day.

Staff should be required to complete a time sheet weekly.

New technology can be used to monitor the whereabouts of leak detection staff. Tracking devices used to improve the security of vehicles can also be used by management to locate the leak detection vehicle during work hours and also to ensure that it is used in accordance with agreed procedures outside of hours. Mobile phone calls can also be traced to the phone network cell if there is any doubt that staff are not where they are supposed to be.

Allocation of work

It is recommended that work should be allocated to staff in manageable packets. A work packet should contain the workload for a week or two ahead. This may be based on a district meter area, or a regular survey zone.

Work packets should contain the relevant technical information required to undertake the task e.g.:

• a plan of the area, showing the DMA boundaries
• a schedule of critical customers supplied in the area e.g. major industrial supplies

- technical data such as pressure regimes, number of properties
- details of any PRVs in the area

An estimate should be made of the time required to complete survey work in a district, and this should be agreed with the leak detection staff.

Equipment audits

Regular inspections should be carried out to ensure that leak detection staff carry the necessary equipment. Electronic equipment should be calibrated at recommended intervals.

Appearance

Staff should be appropriately dressed in the manner set out by the client. Some form of uniform is often worn. High visibility jackets should be worn when carrying out survey work in the highway, and other safety equipment should be used as required. All staff should carry an identity card.

Availability

Checks should be made by mobile phone or radio to ensure that staff are contactable in case of any related operational issue such as a customer complaint.

Progress meetings

Progress meetings should be held monthly between the senior staff responsible for the leakage reduction programme. Meetings between staff on operational issues should be held more frequently. If possible, office space should be allocated to the project as a whole, which can be used by the clients employees and contract staff to build a team spirit and aid good communication.

Abortive excavations

The number of abortive excavations or 'dry holes' should be monitored, ideally with records being kept of the individual or team who located the leak. A suitable procedure should be established to ensure that dry holes are a genuine mistake. If an excavation is made at the point of a located leak and the leak cannot be found, the detection operatives should be called back to the site while the hole is open in an effort to re-check the location. Access to the main at this point may improve the accuracy of a correlation result. Quite often the leak is found within 2 metres of the point of excavation. Other cases have been known of the leak being on the bottom of the main, so it is important to expose its complete circumference.

Out of hours

Detection and repair staff should be contactable outside of normal hours in case they have to be contacted in emergency situations. Problems caused by their valve operations may only be noticed during the normal working day, and water company operations staff may need to consult with them.

Record keeping

Leak detection staff should complete a weekly timesheet. They should also complete an activity report setting out the work carried out and the leaks found. A report should be filed for each leak located, a copy of which will be passed to the repair contractor.

Experienced staff

When contract staff are employed, the client should satisfy themselves of their competence by checking their CVs, and by conducting assessment tests.

A leakage detection contract document should include reference to these issues, and a points allocation system can be used to compare the performance of one contractor or individual against another.

6.7.2 Health and safety

One frequently asked question is whether leakage detection staff should work alone, or whether they should operate in two-person teams. Indeed in some cases there is justification for larger teams. The advantages and disadvantages are summarised in Table 6.2.

In some areas local knowledge of the risk for the safety of staff will dictate the need for two-person teams at certain times of day or night. The best method seems to be to have a mix of two-person teams and single operator working. This allows tasks to be allocated in a cost and time efficient manner.

6.7.3 Managing 'find and fix' contracts

Undertaking leak detection and repair operations requires close communication and cooperation between those responsible for the leak location and those who carry out the repair work. In many cases this can be carried out efficiently by separate parts of the water supply organisation, or by separate contracting companies. However, in some cases the water supplier considers that there are benefits to be gained by having a single contractor carry out the work on a 'find and fix' basis. The perceived advantages are as follows:

Table 6.2 Relative benefits of one- or two-person teams.

Advantages	Disadvantages
Single person working	
More productive time per operator hour	Less control of individual motivation
	Greater risk of an accident resulting in harm to the employee working alone
	More vehicles and equipment per person employed
	Two people are sometimes needed e.g. lifting heavy chamber covers
Two-person teams	
A sense of team work	Not as time efficient for some tasks
Staff monitor each other's activities and safety	
Can be more cost efficient if both team members are gainfully occupied at all times	

- a single point of contact, avoiding split responsibility
- fewer lines of communication leading to more rapid repairs
- lower management and overhead costs

However, there are certain drawbacks which have to be managed if the single contractor system is to be better in practice than having separate parties:

- there may be a tendency to locate and repair leaks which are uneconomic to repair, i.e. the saving in leakage which results is not worth the cost of the detection and repair work
- having a single contractor eliminates the opportunity for comparative performance
- too much control may be placed with the contractor instead of the client

Another option which has been mentioned is to have two companies working together in partnership to gain the benefits but aiming to avoid the complications. There may be a lead contractor and if so it is better for this to be the leak detection company rather than the repair contractor.

6.8   REPAIR TECHNIQUES

This section considers the methods available to effect repairs to leaking mains and service pipes, and the factors to take into account when considering whether they should be replaced rather than repaired. The type of repair will, of course, depend on the type of leak or burst.

6.8.1   Types of leak

The type of repair will depend on the type of leak:

Splits

Some materials such as steel and PVC can split horizontally, usually due to excess pressure, or due to a defect in the pipe wall, or from external damage. In solvent-welded PVC systems, the split can run through joints and may extend for several metres.

Ring crack

The most common type of fracture in cast iron pipes is a vertical ring fracture. Spun iron pipes which have a relatively thin wall and low beam strength are especially prone to this type of failure. Smaller diameter pipes are more prone to ring cracks than larger ones.

Corrosion holes

All metal systems are susceptible to corrosion resulting in holes. In steel mains, the first impact may be a pinhole leak. In cast iron, the wall becomes graphitised, and a piece of the pipe wall may then be blown out under pressure.

Ferrule damage

The point at which service connections are made to the main is a common point of failure. Movement of the service pipe can cause damage, as well as increase the chance of corrosion,

Joint leaks

Joints are another common source of leakage, even in newly laid systems. Pressure testing of new pipelines should always be carried out even on small diameter mains.

Leaks on fittings

All mains fittings which rely on a mechanical joint and a gland arrangement are a potential source of leakage. Fire hydrants, line valves, air valves and flanged joints are a common source of leakage.

6.8.2   Distribution mains

Techniques available to repair water mains include:

Collars

A repair collar is fitted around the pipe without cutting into it. The collar may be cast or ductile iron with some form of rubber seal at each end, and sometimes at intervals along its length. It has two halves which bolt together to form a seal around the outside of the pipe. The collars are available for different outside diameters of pipe, with some limited flexibility on size. Collars are also made from flexible stainless steel with a sheet rubber insert allowing for a greater variation in pipe diameter.

Collar repairs are suitable for pinhole leaks, and vertical fractures where the pipe ends are still in line and there is no damage to the nearest upstream and downstream joint. Types of collars are also available for joint repairs.

Metal collars are also used on PVC systems, and solvent-welded collars are also available. For polyethylene systems, mechanical collars are available, and also fusion-welded collars.

Piece of pipe and couplings

Some repairs require a section of pipe to be cut out of the main. This will be the case if the hole is too large for a collar, if there is a longitudinal split along the pipe (which can happen with some types of plastic system and, in rare cases, with steel pipes), if the ends of a vertical fracture are out of line or if other damage has occurred in the vicinity of the burst.

The repair consists of a short section of pipe (not necessarily the same material as the existing main) and two couplings. The couplings are slotted onto the main at each end, and pulled into place once the new section of pipe has been lined up.

Various types of coupling are available to suit PVC, polyethylene, GRP, ductile iron, steel, and asbestos cement pipes. Some couplings are suitable for a variety of materials and have flexibility on the outside diameter they will fit. Using them results in fewer items required to be kept in stock.

Wooden plugs

Repairs of pinhole leaks on steel mains can be effected by hammering in a pointed wooden 'plug', often while the pipeline is still pressurised. A steel patch can then be welded over the top of the plug. Wooden plugs can also be used to make temporary repairs to other types of water mains

6.8.3   Transmission mains

Repair techniques for transmission mains are similar to distribution mains. However, due to the larger diameter giving higher beam strength, the there tends to be a lower proportion of vertical fractures. Transmission main repairs usually involve joint leakage, pin holing, or leakage from fittings. Splits can result in spectacular bursts but they are relatively rare. If they start to become common then replacement of the pipeline should be considered due to the major impact they have.

6.8.4   Service pipes

Service pipe repairs include collars, or replacing a section of pipe and couplings. Often the repair is associated with a defective stop tap, or meter, or ferrule connection, rather than the pipe material itself.

Consideration should be given to a policy of replacing, rather than repairing, service pipes once they leak. The difference in cost between repair and replace for service pipes is much less significant than that for water mains.

6.8.5   Replace or repair?

At some stage, the water supplier has to decide whether to continue carrying out repairs on the mains in an area, or whether they have they have reached the end of their serviceable life and should therefore be replaced. There are a number of factors which support this decision:

- the relative costs of continuing to make repairs versus the cost of replacement
- the impact on customer service of continuing to accept the interruptions to supply caused by bursts and the subsequent repairs
- the requirement to meet set targets for leakage, which will be assisted by the installation if a new main
- the finance and other resources available

It is very difficult to cost justify replacing mains solely on the grounds of leakage reduction unless they have exceptionally high burst rates, or high background leakage, or the value of the water in the are is so great that costly capital works can be shown to be economic. However, mains replacement gives other benefits in terms of improvements to water quality and improved customer service levels. Taking these benefits into account may result in a positive justification for replacement.

6.9   LEAK DETECTION IN NETWORKS WITH INTERMITTENT SUPPLY

This section addresses the problem of leak detection in networks with intermittent supply, where conventional monitoring techniques and acoustic detection techniques are not possible. In India, where demand is usually far in excess of available supplies, most of the systems only supply water intermittently. In some systems production of water can barely meet the minimum demand of the consumers, and excessive leakage is an aggravating factor. The problem is exacerbated by the following:

- duration of supply has to be restricted to reduce the high volume of leakage and customer waste
- longer periods of supply between interruptions reduces peak demand, but this is feasible only if measures are taken to restrict leakage and the overall system capacity
- shorter periods of supply between interruptions subjects the distribution system to high flowrates and low pressure
- customers tend to over-ride the mains supply and draw as much water as possible by building underground tanks/sumps and pumping water into the house
- water operators tend to lay additional mains in order to improve pressures in the system, which results in an overall increase in system capacity
- supply at low pressure and for short duration restricts leakage in a system, but this is unsatisfactory for the customer, and causes a public health risk from infiltration
- conventional leakage monitoring and detection methods (e.g the minimum night flow method) and pressure management to reduce leakage, are not applicable to intermittent supplies – a reasonably high mains pressure is required for leak detection equipment to be used effectively.

6.9.1   Diversion of supply

The *stop tap* method has been successfully used in several cities and towns in India to quantify leakage and to locate the leak points. The method requires the temporary diversion of supplies to a test area The steps are as follows:

1 The test area is isolated by closing boundary valves.
2 Stop taps on customers' service connections are closed.
3 A special supply is arranged for the test area from the nearest water distribution station.
4 Water meters are used to measure the flow through the test area. This is a direct measure of leakage.
5 Leak detection equipment is used to locate leak points.

The disadvantages of this method are:

- arrangements have to be made for supplies to be diverted to the test area
- a considerable volume of water is lost from leak points during the test
- supplies to adjacent areas are affected, leading to customer complaints

- not all leaks can be identified in a short test and repeat tests are required
- it is labour intensive

6.9.2   Mobile tanker method

To overcome the major disadvantage of diverting supplies the mobile tanker method has been developed. Water mains are charged over a timed duration, long enough to measure leak flow and to use selected equipment to locate leaks. The approach was first tested by Tata Consulting Engineers [9] in Ahmedabad city (population 2.5 million) in central India and has been extensively used in Madras Metropolis (population 4 million).

The method has the following advantages:

- it has been specially designed for intermittent supply systems
- there is no disturbance of normal supply
  - only a small area is isolated
  - only a small volume of water is used for testing.

The mobile tanker unit consists of:

- a street water tanker
- a wheel mounted pump
- an easily made pipe assembly with valves to control pressure
- a turbine water meter unit with pulse head and data loggers.

A small test area is isolated. Water is drawn from the tanker and is injected into the area using the mobile pump. A bypass pipeline returns the water partially to the tanker. Valves are provided on the pump delivery return lines. By manipulating these valves, the desired pressure is maintained. The line supplying the injection point contains a meter with a pulse head and a data logger for recording the flow. A pressure transducer is also provided to log the pressure at the injection point. Loggers are downloaded onto computer and graphs of flow and pressure with time are obtained.

Customer support is ensured by distributing leaflets. This develops their awareness of the test and builds confidence in future improvement of their supplies.

A typical test area is 100 connections or 500 m of pipe length. Boundary isolating valves are closed and stop taps at customer premises are closed (or the service pipes cut and plugged). In areas where operating staff are reluctant to operate valves which have been throttled to adjust supplies, mains can be cut and capped.

The test is carried out during non-supply timings so there is no inconvenience to the consumers. The test sequence is as follows:

1 Water is injected into the isolated test area through mobile pump tanker unit.
2 A prescribed pressure at the injection point is maintained by manipulating the unit valves.
3 As the area is isolated and the house services are plugged, the reading on the in-line meter is a direct indication of leakage in the test area.
4 An assessment of leakage can be made for a range of pressures which would prevail in the distribution system later.
5 Pressure transducers and data loggers are also introduced at the other isolation points, to check whether the entire area is pressurised to the desired level.
6 Once the leak points are rectified, it is also possible to assess the reduction of leakage in the test area by repeating the test.
7 The improvement of pressure at the isolation points of the mains is an indicator of the overall improvement due to leakage reduction.
8 The exercise is completed when the entire stretch has been scanned by leak detection equipment, leaks pinpointed and the service connections and mains restored.
9 In Madras, the total duration for this exercise beginning with segregation of area, carrying out test and returning to the normal system for supply took about 8 hours.

A sample survey helps to identify the operating practices and the need to alter them or correct deficiencies. Examples are:

• missing ferrules at house connections
• service connections made with insufficient cover/protection
• poor repair practices (e.g. bicycle tyre rubber)

In some systems leakage is so high that efforts to pressurise a sub-district uniformly are at first unsuccessful. Only after repairs have been carried out in part of the test area does the pressure rise to a suitable level for the test.

6.10   REFERENCES

1 Farley, M (1985) *District Metering Part 1: System Design and Installation.* WRc Report ER 180E. Swindon: WRc.
2 Farley, M and Merrifield, T (1987) *District Metering Part 2: System Operation.* WRc Report ER 210E. Swindon: WRc.
3 WSA/WCA Engineering and Operations Committee (1994) *Managing Leakage: UK Water Industry Managing Leakage* Reports A–J: Report A – *Summary Report*; Report B – *Reporting Comparative Leakage Performance*; Report C – *Setting Economic Leakage Targets*; Report D - *Estimating Unmeasured Water Delivered*; Report E – *Interpreting Measured Night Flows*; Report F – *Using Night Flow Data*; Report G – *Managing Water*

Pressure; Report H – *Dealing With Customers' Leakage*; Report J – *Techniques, Technology and Training*.London: WRc/WSA/WCA.

4   UK Water Industry Research Ltd. (1999) *A Manual of DMA Practice*. London: UK Water Industry Research.

5   Technical Group on Waste of Water (1985 [1980]) *Leakage Control Policy and Practice*, Standing Technical Committee Report no. 26. Original publication London: Doe/NWC. Reprinted London: WAA/WRc.

6   Farley, M (1997) 'Innovative ways to detect leakage'. Paper presented at IIR Conference 'Water Pipelines and Network Management', London, 17–18 February.

7   Trow, S (1999) 'Outsourcing leakage management: developing contractual terms to ensure accountability'. Paper presented at IIR Conference 'Water Leakage', London, 21–22 September. London: IIR.

8   Pook, G (1997) 'Contracting leak detection: payment by results'. Paper presented at South West Water seminar 'A Leakage and Network Strategy for the Millennium', London, 12 February. London: SWWS/Malcolm Farley Associates.

9   Kumar, A (Tata Consulting Engineers) (1991) 'Leak detection using the tanker technique'. Paper presented at Workshop on Leakage Detection and Prevention of Water Supply Systems, Surabaya, 2–5 December.

7

Pressure management

7.1   INTRODUCTION

The rate of leakage in water distribution systems is a function of the pressure applied by pumps or by gravity head. There is a physical relationship between leakage flow rate and pressure, which has been proven by laboratory tests and by tests on underground systems. It is simple to demonstrate this principle:

Take a bottle or a bucket, or some other suitable container. A 2-litre plastic soft drinks bottle works well. Fill the container with water. Make a hole in the side near the bottom. The water will spurt out. Now make another hole half way up. The flow rate will be less. Now make a hole near the water surface, the water trickles out. The same principle applies in water distribution systems, the pressure being equivalent to the depth of water above the hole.

Burst rates are also a function of pressure. The strength of the relationship, and the quantification of it, is not as well understood as the relationship between flow rate and pressure. However, there is still considerable evidence to show that burst frequency is very sensitive to changes in pressure. Evidence shows that the rate of increase of bursts is more than linearly proportional to pressure. Indeed it has even been suggested that there could be a cubic relationship i.e. burst frequency proportional to pressure cubed.

© 2003 IWA Publishing. *Losses in Water Distribution Networks* by Malcolm Farley and Stuart Trow. ISBN: 1 900222 11 6

Pressure management is one of the fundamental elements of a well-organised leakage management strategy. It should be an integral part of the strategy because it impacts on several other aspects:

- The rate of rise of leakage is a function of pressure. If pressure is reduced, the rate of increase in leakage will reduce. Therefore, there is an impact on the level of leak detection resources required.
- The flow rate from all leakage paths (bursts and background leaks) will reduce, so reducing total losses
- The data used to calculate leakage targets and economic levels of leakage should be revised when pressure management is introduced
- Reducing pressure may make existing leaks more difficult to find, because they make less noise, or do not come up to the surface. Therefore, pressure reduction should be coordinated with leakage detection and repair operations.

Pressure management is best undertaken in conjunction with district metering, or establishing supply zones. Therefore, segmenting a system should be considered at the outset of any project, rather than later.

As well as the reduction in pressure, good pressure management will also result in more stable pressures, causing less strain on the pipe network, and less chance of fatigue damage at joints.

Water suppliers who understand the importance of pressure management and take appropriate steps to control pressures, will reap the benefits over a long period of time. These benefits can be set against the initial cost and the cost of ongoing maintenance.

As with all aspects of leakage management, the economics of pressure management follow a law of diminishing returns. However, if the investment is carried out in the most cost-effective way, and the future maintenance costs are minimised by giving due consideration to servicing, spare parts, call out charges etc., then extensive pressure management can be justified in the majority of systems.

7.2   BENEFITS OF PRESSURE MANAGEMENT

This section sets out the benefits of pressure management, the factors affecting the economics of pressure management, the policy issues for the organisation, and the considerations for design, installation, and maintenance of pressure management systems.

There are several benefits of pressure management, and if it is designed and maintained well, there are few, if any, disadvantages.

Reduction of leakage – bursts and background

Reducing pressure will reduce the flow rate from all background leaks and all bursts. The relationship between pressure and flow rate is described in more detail in section 7.5.

Reduction of pressure-related consumption

Reducing pressure can reduce some types of consumption. Any consumption from devices connected direct to mains pressure will give a reduced flow rate at reduced pressure. Examples include taps, showers, and hose pipes. WCs and urinals, which use a flush valve rather than a cistern, will show a reduced consumption. With un-vented boiler water systems driven by mains pressure, the effect will be experienced on the hot water system as well as the cold water.

It is thought that the tendency to leave the tap running for longer at lower flow rate is more than compensated by the reduced flow rate, so that the overall volume used is lower.

If the device is connected to a header cistern (perhaps in the loft space) there will be no impact from pressure reduction. Therefore, it is important to understand the predominant type of plumbing system in an area when predicting the effect of pressure management on consumption.

It may be thought that reducing pressure would have no impact on WC use, because they give a fixed volume per flush. However, there are two effects which cause consumption to be reduced. The first is the effect on the flow of water which runs through the cistern and into the pan when the WC is flushed. Some of the water which goes down the pan is from the stored water in the cistern. Another quantity enters the cistern and runs straight through it. In lower pressure areas, this flow rate is lower, and so the flush volume tends to be lower.

The second effect is due to the sealing of the ball valve. In higher pressure systems there is more of a tendency for the ball valve to drip, so increasing the stored volume between flushes, and the tendency for the cistern to overflow. When the ball valve does shut off, it tends to be at a higher level than for a low pressure system. With lower pressure there is more chance of the ball valve seating and shutting off the flow rate.

It must be stressed that these are phenomena which are not experienced in each and every WC cistern. The impact depends on several factors including the age and type of the cistern and ball valve, and the layout of the internal plumbing system. In some cases, lower pressure will result in increased leakage at the ball valve seat, depending on the seating arrangement.

If a major pressure management investment is dependent on the impact of reduced consumption, then it is recommended that tests be carried out to assess the impact in designated trial areas.

Reduction of frequency of bursts

Data from one UK water supply company shows the reduction in burst frequency before and after pressure management was installed in an area. The data set is limited in size, but it indicates that a unit reduction in pressure will give a 3 or 4 times reduction in burst frequency e.g. reducing pressure from 80 m to 40 m (a 2:1 reduction) will reduce the burst rate from 7 bursts per 100 properties per year to only 1. Of course there are many other factors which affect the burst frequency of mains including:

• weather conditions
• pressure surges
• accidental damage
• ground movement
• corrosion

Therefore, it is difficult to obtain good quality data to prove the strength of the relationship.

Burst frequencies will be more reliable in larger areas, e.g. supply zone, but at that scale it is more difficult to make significant changes in pressure. Therefore most data is available at DMA level, where the burst rate is more erratic, and so it may take several years to determine the true benefits.

Figure 7.1 shows the relationship between average zone night pressure (ANZP) and burst frequency for a sample of data from one UK water company [1].

Figure 7.2 uses data from a second UK water company [1], expressed in terms of bursts/1000 km/year, to show a similar relationship, with approximately a 4:1 factor.

The form of the relationship below 40 m night pressure is difficult to judge. In the UK, regulated standards of service prevent pressure being reduced below about 15 m at any time of day. Therefore, to allow for head losses in the network, and for variations of flow demand on different days of the week and at different times of year, pressure-reducing valves (PRVs) are set to give no less than about 20–25 m at peak demand time. Therefore night pressures are rarely below 30–40 m, unless some form of flow modulation is in place.

Provision of more constant supply to customers

Without pressure management, the pressure at customers' premises will be a function of the pressure of water where it enters the distribution system, less the head loss through the underground pipe network.

The pressure on entry will depend on whether the water is pumped direct into the mains from boreholes, low level service reservoirs and tanks, or whether it gravitates from high level service reservoirs, spring collection tanks etc. Head losses

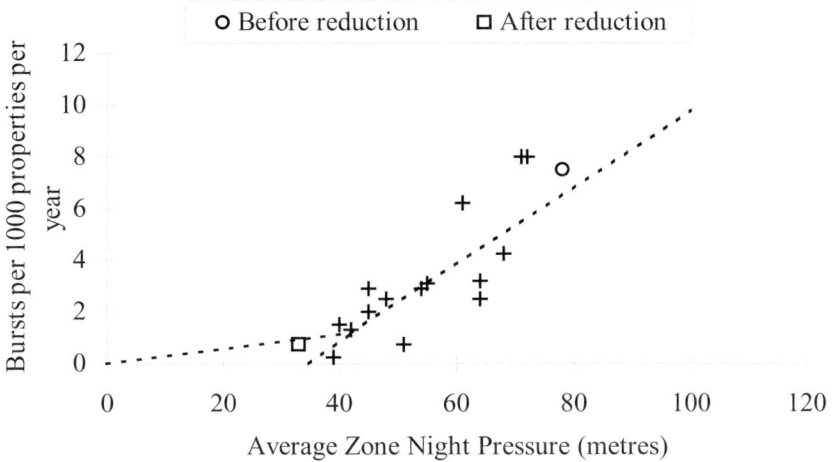

Figure 7.1 Relationship between average zone night pressure (ANZP) and burst frequency for a sample of data from one UK company based on DMAs in an urban area.

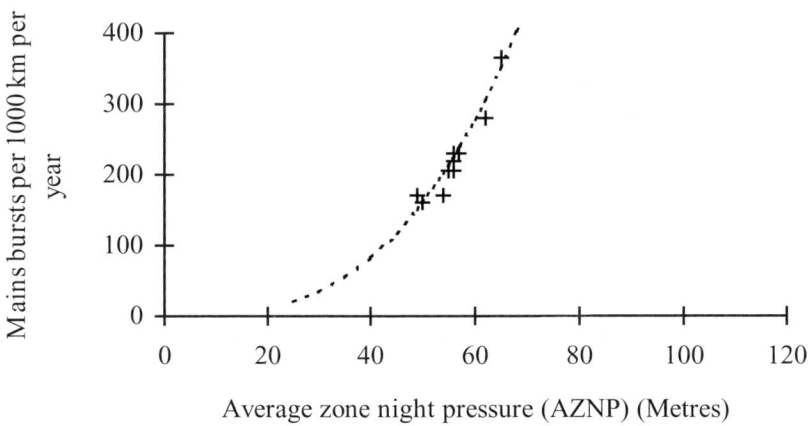

Figure 7.2 Relationship between average zone night pressure (ANZP) and burst frequency for a sample of data from a second UK water company based on large supply zones (average 100 000 properties).

in the distribution system will depend on the capacity and configuration of the system, the fluctuation of demand over the day, and the internal condition of the mains.

Taking all these factors into account, the pressure at customer premises can vary considerably between day and night and on different days of the week. Without pressure management, there is an increased risk that pressure at customer premises will fall outside acceptable limits. Low pressure will result in a greater number of 'no water' complaints, or complaints that appliances such as mains fed water heaters will not operate. Pressure which is too high may result in reverberation of pipes, difficulty in maintaining correct temperature on shower mixer valves, and cavitation damage on taps and mixers.

Well-organised pressure management regimes will result in a better understanding of the factors affecting pressure at customer premises, and will allow systems to be put in place to maintain pressures within specified bounds.

In systems which are not charged 24 hours a day, (intermittent supply systems) it may be possible to reduce pressure while the system is charged in order to reduce the leakage rate at the time. The reduced leakage will result in more water being available so that the time for which water is available can be extended.

Increased fire fighting capability

In a similar way, lack of pressure management may result in inadequate supplies from fire hydrants for fire fighting. Many water supply organisations avoid pressure management because they are concerned that it will reduce availability of fire fighting water, and result in disputes with the fire authorities. However, with modern technology and design techniques it is possible to reduce pressure (and therefore leakage) and also provide adequate supplies for fire fighting.

The important issue here is the difference between pressure at fire hydrants and the flow available when they are opened up. It is not true that high pressure will always give high flow. If the hydrant is connected to a small diameter main, or one which has restricted capacity (perhaps due to internal corrosion), it will give a lower flow than one which is connected to a larger main with higher capacity, even though pressure in the main may be lower.

Some fire authorities focus only on the static pressure, whereas in others regular tests are carried out to prove the flow available. In some countries fire tenders carry their own water supply, sufficient to begin fighting a fire, and the hydrants are used only to re-fill the tanks on the tender.

Protection of the long-term life of the assets

Daily variations in pressure place stresses and strains on the pipe network. Damage can occur at joints and fittings, and pressure splits can occur in some types of pipe.

Damage may be due to a fatigue effect, which takes place over a long period of time. The greater the range of pressure variation, and the more frequent the variation, the greater the chance of damage caused by pressure fluctuations.

Pressure management will tend to smooth out these variations, resulting in less damage to the network, and a longer asset life expectancy.

7.3   POTENTIAL PROBLEMS

There are few problems with pressure management, provided the systems are designed, installed and maintained correctly. Further details are given in sections 7.8–7.13.

It must be understood however, that as with all forms of investment, there is a maintenance requirement. If systems are not maintained, equipment such as pressure-reducing valves will not operate effectively, and pressures will fall or rise outside of the set limits. In extreme cases, the valve may malfunction and become unstable leading to the possibility of burst mains or inadequate supplies to customers.

7.4   THE RELATIVE IMPORTANCE OF PRESSURE MANAGEMENT

Taking into account all the factors which govern the level of leakage and losses in water distribution systems, it is possible to compare the cost-effectiveness of the various measures available for reducing leakage and losses. The relative ranking of these factors will depend on the individual systems and the amount of work which has been carried out already. For example, pressure management will be less cost-effective in a flat area than in an area where the ground level has significant undulations. Pressure management will be less cost-effective when the 'easy to do' schemes have already been carried out, than if there has been no pressure management at all. Therefore, care should be taken when applying generalisations to specific supply zones.

However, it is possible to make a general comparison based on experience from leakage reduction projects around the world. The most important factor tends to be the condition of the infrastructure, and its propensity to burst and leak. The condition will be a function of:

- the age of the system
- the type of pipe material
- the method of jointing the pipes
- the ground conditions
- the surface loading

Table 7.1 Leakage rate (volume per unit time) varies with pressure to power N_1.

Country	Number of sectors tested	Mean value of N	Range of values of N
UK (1977)	17	1.13	0.70 to 1.68
Japan (1979)	20	1.15	0.63 to 2.12
Brazil (1998)	13	1.15	0.52 to 2.79

Infrastructure condition is inherited from previous owners or managers of the network, and significant improvements take a great deal of time and money. Therefore, changing infrastructure condition to reduce leakage is a long-term option at best, and in many cases the improvements which can be justified economically have a small impact on leakage levels compared with other options.

Pressure tends to be the second most important factor in determining leakage level, after infrastructure condition, but pressure management is more cost-effective than infrastructure management. In the majority of cases, pressure management forms a key element of the leakage reduction strategy.

7.5   PRESSURE/LEAKAGE RELATIONSHIPS

Most hydraulics text books refer to the relationship between pressure and the flow through an orifice. Splits and holes in pressurised pipes causing leakage will act as orifices. The relationship is stated as:

$$V \quad C_d \sqrt{2gP}$$

where V is the velocity of water through the orifice in m/sec
C_d is a discharge coefficient: a dimensionless factor of less than 1
g is the gravitational constant in m/sec^2
P is the pressure in metres head

So, for a hole of a given area, the flow rate in m^3/sec varies with the square root of the pressure, i.e. leakage is proportional to $P^{0.5}$. This relationship can be proved on laboratory test rigs.

In practice, when measurements are taken from district meter areas, the relationship tends to be more pronounced. The reduction in leakage from pressure reduction is more than predicted from the theoretical relationship.

Tests carried out in Japan, the UK and Brazil between 1977 and 1997 [1] show that, on average, leakage varied with pressure to the power of 1.15 not 0.5. These results are summarised in Table 7.1.

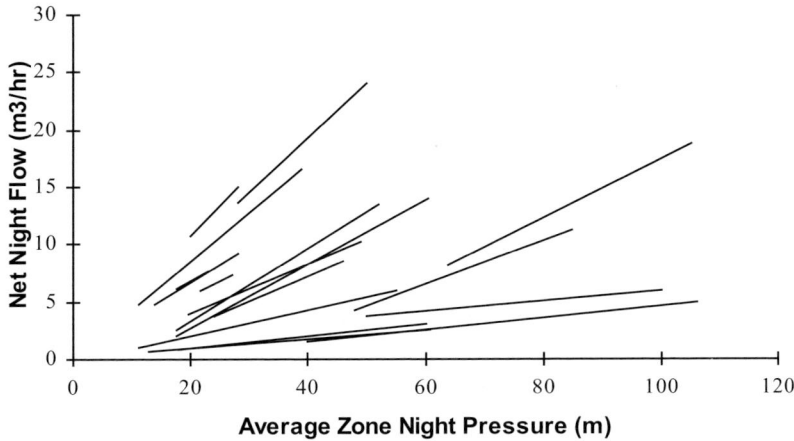

Figure 7.3 Impact of changing AZNP on net night flow.

There are other examples of tests being carried out [2] from which a general empirical relationship has been developed, and attempts have been made to fit an equation to the empirical data. The results are similar with some variations depending on the way in which the tests were conducted, and whether the relationship is between pressure and leakage, or pressure and flow or night flow.

However, while the aggregation of tests in individual districts tends to be similar, there are major differences between the individual tests themselves. This is illustrated in Figure 7.3.

Analysis shows that the power factor (N_1) can vary between values of about 0.5 and values over 2. This is difficult to explain from the test rig results.

In 1994, a new theory was proposed [3] – as well as the flow velocity being a function of pressure, perhaps the area of the orifice varied with pressure in some situations. This would explain the variability between one district and another. A district with leakage predominantly from fixed area holes (e.g. corrosion pin holes in metal pipes) would tend to have N_1 values of about 0.5. In districts where the holes vary in size proportionately with pressure the N_1 value will tend towards 1.5 (the area varies with P^1, and the velocity varies with $P^{0.5}$, so together the flow rate varies to $P^{1.5}$). Values greater than 1.5 are explained by the existence of leakage paths which increase in size in two directions, so they vary with P^2.

When assessing the cost–benefit of pressure management, it is clearly important to understand the effectiveness of the proposed schemes in the area under consideration. However, due to the variability this can be difficult, Therefore the following general guidance is offered:

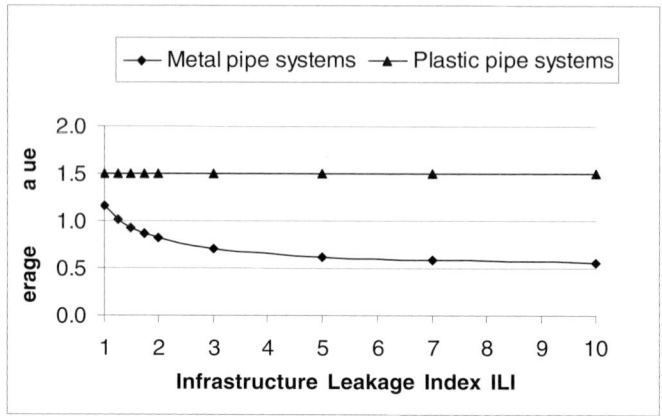

Figure 7.4 N_1 as a function of infrastructure leakage index.

- When considering pressure management as a general policy or for a relatively large supply zone, it is reasonable to assume a linear relationship between pressure and leakage flow rate (i.e. $N_1 = 1$). This is due to the aggregation.
- When assessing the effectiveness of pressure reduction in an individual district, an assessment can be made of the N_1 value from two factors, the predominant materials of services and mains, and the initial leakage level.

Figure 7.4 shows the predicted relationship between the infrastructure leakage index (see Chapter 3) and the N_1 value [4].

In systems which are comprised completely of plastic mains and service pipes, then the N_1 value will tend towards 1.5, regardless of the level of leakage. In metal pipe systems, the N_1 value is a little over 1 for systems with very low leakage, with normal N_1 value between 0.75 and 1. As leakage increases due to the predominance of bursts, rather than background leakage, the N_1 value tends toward 0.5. If there is insufficient information to make such an assessment, then an average of 1.15 is reasonable.

If the pressure–leakage relationship is critical to the accuracy of the result of some other exercise, then a test should be made in the specific district to alter the pressure and measure the impact on leakage level based on the nightflow data. The flow and pressure data can be analysed to determine the N_1 value [4], as shown in Figure 7.5. This approach may be required in control areas used to determine the policy on pressure and leakage management, or in districts used to assess per capita use in unmeasured properties.

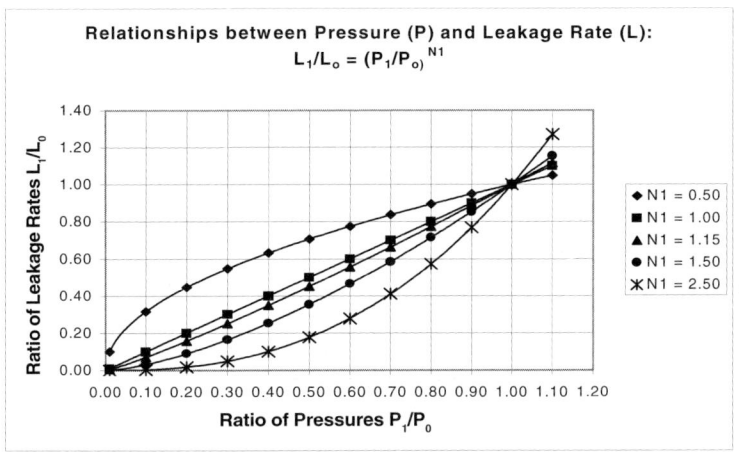

Figure 7.5 Relationship between pressure and leakage rate.

7.6   ECONOMICS AND COST– BENEFIT ANALYSIS

Cost–benefit analysis of pressure management can be carried out at a number of levels. For general company policy and target setting, it is likely that assumptions will have to be made using experiences of other companies initially. As work is carried out the results will be used to review the validity of these assumptions. For a supply zone, an assessment can be made of the reduction in leakage by reducing the average pressure in the zone, using a component-based leakage model. The cost of reducing the average pressure can be assessed in a number of ways e.g. by experience of the cost of a similar scheme for another similar zone, by assessing the likely cost per property metre (see section 7.6.2), or by carrying out the initial designs of all potential schemes. For an individual district, the actual costs of the proposed scheme will be compared to the forecast benefits.

The purpose of this section is to set out the factors which have to be considered when deciding on the economic extent of pressure management, and when considering whether it is cost-effective to proceed with a particular scheme.

7.6.1   Costs

When considering pressure management, there are a number of costs to be taken into account. It is important to understand whether these costs are one-off items, or continuing expenditure. It is recommended that a schedule be prepared setting out each cost under the following headings, and projecting the cost over say a 20-year time period. Some costs will occur in the first year only. Others will occur year after year, while some may occur at intervals e.g. PRVs may need to be replaced every few years.

Design

The design cost will include the following elements:

- data logging to understand the pressure regime within the area to be managed
- analysis of data
- dealing with customer issues
- field test (for newly created areas)
- preparation of reports

In cases where the design is carried out by consultants, it is possible to obtain a cost per scheme for the design work, rather than having to estimate a cost based on a time and fee rate basis.

Purchase of equipment

There is the cost of the pressure management equipment itself i.e. PRVs and flow controllers, to consider, but this is usually small compared to the additional cost of line valves, fittings, chambers and covers. Therefore, it is usually false economy to select the lowest-cost PRV and controller. The whole life cost, including maintenance and replacement charges, and the costs of dealing with incidents due to malfunction should also be considered. Figure 7.6 shows a typical globe type valve used for pressure management.

Installation

This will include excavation, and installation of the pipework. Consideration should be given to optimising the design to consider whether:

- the PRV should be on a by-pass or on-line with a by-pass around it
- the PRV should be installed in the same chamber as a district meter

Commissioning

Commissioning is usually inexpensive, and can be included in the cost of installation. However, some organisations prefer to commission a pressure management scheme in stages over a number of weeks. Therefore, a number of repeat visits will be required which may mean that the cost has to be included. Special schemes, or those involving a high degree of customer contact may also be expensive.

Figure 7.6 A typical globe type valve used for pressure management.

Monitoring

The installation should include tapping points, either on the valve itself or on the pipework, to allow the upstream and downstream pressures to be monitored. Monitoring should either be carried out at regular intervals (logging at least once a year is recommended) or provision should be made for continuous monitoring via some form of telemetry system. On no account should a pressure management scheme be installed and left alone indefinitely.

Valve maintenance

Different makes of PRV require different levels of maintenance. If only a small number of valves are installed, and there is a maintenance operative available to carry out the work, then this is not a major issue. However, if there are a significant number of valves in operation, or if there is insufficient resource to deal with maintenance as part of other duties, then the cost implication will dictate that maintenance should be kept to an economic minimum. Maintenance will include:

- cleaning of filters on the control pilot loops
- cleaning debris from inside the valve which can be become lodged in the internal mechanism
- greasing of moving parts
- replacement of worn seals
- replacement of worn parts, or torn diaphragms

Careful choice of the valve will avoid many of these maintenance issues.

Boundary maintenance

As with the establishment of DMAs, this is a major issue. Careful planning and design will overcome the need to open up the boundary of a pressure managed area for routine operations. However, there will be occasions, such as a burst main, when it is necessary to open a boundary valve to back feed supplies from an adjacent area.

From this assessment of costs, the net present value (NPV) of the cost stream can be estimated, and this can be compared to the NPV of the benefits to determine whether the scheme is cost effective. The NPV of the benefits will include savings due to reduced leakage, reduced burst frequencies, improved asset life, and reduced customer contact. More details of the benefits are given in section 7.2.

7.6.2 Cost per property-metre (pm)

A simple method of comparing the cost-effectiveness of alternative pressure management schemes is to consider the cost per property-metre. The cost is the net present cost derived as set out above. The benefit can be estimated in terms of the reduction in pressure (in metres head of water) multiplied by the number of properties (or perhaps the length of main) to be included in the pressure management scheme. The denominator is known as the property-metre. For example:

- district of 1000 properties
- pressure reduced by 15 m average = 15 000 pm
- cost of scheme = £7500 (€/US$12 375)

    Cost per property-metre = 7500/15 000 = £0.50/pm (€/US$0.825/pm)

Analysis of schemes in the UK has shown that pressure management can be cost justified up to about £3–5 / pm (€/US$5–8 / pm). If the cost exceeds £5/pm then it is difficult to justify. If the cost is between £2 and £5, then a more detailed cost–benefit analysis is worthwhile. If the cost is below £2/pm, then there is justification in proceeding with the scheme without further detailed analysis.

Figure 7.7 shows how the cost per property-metre can be used to prioritise potential pressure management schemes, according to their cost effectiveness.

7.7 POLICY ISSUES

When considering pressure management, there are certain policy issues which have to be included in the strategy. These are key issues for the organisation which should be addressed by senior staff in advance of investment in pressure management. Otherwise, they will be addressed by technical staff as work is carried out – decisions

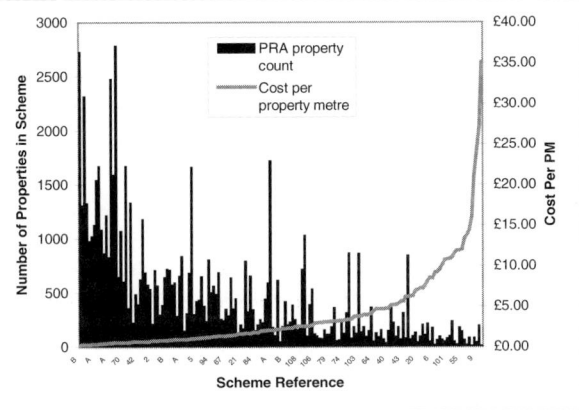

Figure 7.7 Potential pressure management schemes for an area ranked by the cost per property-metre.

will then be taken at a local level which may contradict others taken in another part of the organisation. Faced with customer problems, incorrect advice or information could be given, and the pressure management scheme may be compromised as a result of changes being made in haste.

Levels of service

When reducing pressure to reduce leakage or reduce burst frequency, there is a fundamental question which has to be addressed: 'What is the minimum acceptable pressure which has to be maintained?'

The question has several strands:

- Where do we measure the minimum level of service?
- For how long each day can we accept a pressure below this level?
- On how many days of the week can we accept a pressure below this level?
- Are there certain seasons in the year when a higher level is required?
- Is this a mandatory, guaranteed or regulated standard, or is it an internal policy only?

Different standards of service may apply to domestic customers and commercial or industrial customers, or there may be one single standard. In some cases, industrial customers are willing to pay for a higher standard of service to ensure water is available for processes or for fire fighting. Sprinkler supplies have different grades, and it is common for commercial customers to pay the capital cost of installing

Table 7.2 Customer notification.

Period of notice	Customer group	Notification method
6 months	Industrial	written
	Commercial	written
	Hospitals	written
	Renal dialysis etc	written
	Fire Service	written
3 months	Non-customer e.g. councils, press	written
2 months	Domestic local	newspaper
6 weeks	Fire Service (final details)	written

additional mains capacity to ensure sufficient flow is available. This may be paid as a single capital contribution or as part of the annual cost. Charging for these mains places an additional obligation on the water supplier. Maintaining these obligations can constrain the scope for pressure management.

Customer notification

One of the more difficult strategy questions is whether or not to inform customers of a proposed change to pressures. This can be a dilemma, and there is no correct answer. If customers are informed then they are more likely to object – this will at best delay progress, and at worst prevent schemes from going ahead. On the other hand, if customers are not told in advance, and they subsequently experience difficulties with lower pressure, or from interruptions and dirty water caused by the installation work, then they will naturally complain.

The answer therefore seems to be to tell some customers something, but not to tell all customers everything. The first priority is to meet any statutory or contractual commitments. Then there is a choice to be made. Table 7.2 is based on *Managing Leakage* Report G [5]. It will not be applicable everywhere, but it can be varied to suit local circumstances. The important thing is that the issues are fully considered, and that a policy is developed which is appropriate for a particular organisation at a particular time.

7.8   DESIGN OF PRESSURE MANAGEMENT SCHEMES

Pressure management is not just about installing pressure-reducing valves in conjunction with district metering or at source works. It is important to consider the whole water supply and distribution system from source to tap. Pressures should be optimised in an integrated long-term strategy, and pressure should be an ongoing consideration – in many aspects of the operation, maintenance, and rehabilitation and extension of the pipe network. It should not be a one-off project, managed by a small number of staff, or by external consultants. It should be a consideration of all staff involved with the system. Pressure management should be carefully planned and carried out in a coordinated way. It should not be done in a haphazard way, with schemes being carried out in an ad-hoc manner without due consideration for the knock on effects. For example:

- When designing district meter areas the scope for pressure management should be evaluated in all cases. A cost–benefit analysis should be carried out, for fixed outlet PRVs and for flow-modulated PRVs (see section 7.10).
- When responding to customer complaints of either low or high pressure, some analysis should be made to understand whether the pressures in the system are at an optimum.
- Every outlet of a service reservoir or tank or pumping station has the potential to be pressure controlled. A relatively straightforward analysis, at little cost, will show whether there is scope for pressure reduction. If there is scope, the case can be made to carry out a study to quantify the benefits and the costs, and the operational consequences.
- Every branch off a trunk main offers a potential pressure management scheme.
- When changing, refurbishing, or extending the system (e.g. to meet the demands from new housing or industry), care should be taken to maintain the integrity of any existing pressure management scheme, and to take any available opportunity for further pressure management.
- Even new systems leak, so there should be a design standard for maximum and minimum pressures, which should be taken into account when changes are made to the network.

The initial analysis of potential areas for pressure management will utilise a number of sources of information such as:

- network model results
- local knowledge of operations staff
- customer contact files
- GIS records

Figure 7.8 illustrates the concept of a pressure-managed area (PMA)

7.8.1   Types of pressure management scheme

Pressure management comprises a combination of different techniques. All of them should be included in a well-developed programme of action to gain control of the

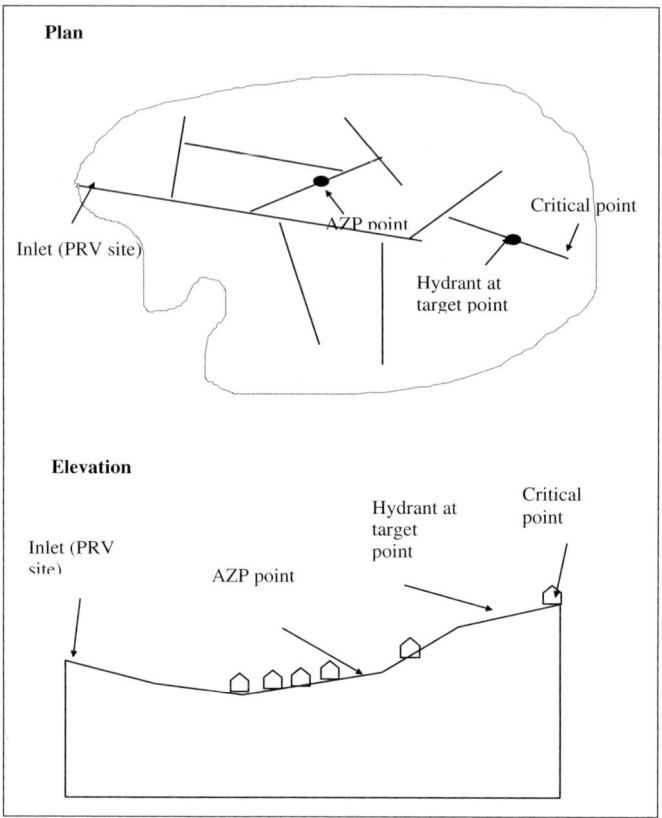

Figure 7.8 A schematic diagram of a pressure-managed area (PMA).

water distribution system, in order to achieve the benefits set out above. This section outlines some of the options available:

Install pressure-reducing valves (PRVs)

PRVs are designed to create a head loss which reduces pressure on the outlet side. There are several variations in design:

- globe type diaphragm-actuated valves
- rolling diaphragm valves
- direct-acting spring-loaded valves
- weight-loaded valves
- electrically-controlled valves

Some valves create a fixed ratio reduction in pressure, e.g. 3 to 1, regardless of flow rate. Others can be fitted with pilot valves, which alter the position of the valve such that it gives a fixed outlet pressure regardless of inlet pressure or flow rate.

The type of valve will depend on its function in the network. Direct acting spring-loaded valves tend to be used on small-diameter service connections. Diaphragm valves are popular for use in conjunction with district metering. Electrically-controlled valves tend to be used on trunk mains to control flow, often by telemetry signal, between treatment works, pumping stations and service reservoirs.

Introduce a break tank

A break tank can be installed in a main to break the hydraulic gradient at a convenient point, and drop the pressure to atmospheric. This is a convenient and reliable method of controlling pressure on long lengths of trunk main across undulating terrain. In some cases, the water discharges into a break tank and is then pumped into the next section.Where storage is required, a service reservoir will have the same effect.

Rezoning

By installing link mains and line valves, it is possible to supply areas in different ways. This could allow areas to be transferred to adjacent lower pressure systems. Alternatively, some properties in an area may be transferred to a higher pressure supply to allow the remainder of the area to be pressure reduced e.g. if there is a mixture of high and low rise buildings in the area.

Trunk main control

The installation of pressure-reducing valves, or electrically-controlled valves on trunk main systems can be used to control pressures over a widespread area of supply. This usually involves relatively small pressure reductions at night, but because of the large area, the benefits are high compared with the localised cost.

Combined booster and PRV

Areas of high ground may be boosted by an in-line pumping installation so that the remaining lower-lying area can be pressure reduced. The cost of pumping to a small area is offset by the savings from reduced leakage and reduced mains failures in the larger area.

Pump control

Slow-start controls can be worthwhile on pumping sets to prevent pressure surges. Pressure relief valves, which bleed water back to the inlet side of the pump can prevent pressures exceeding prescribed levels. As pump output reduces, there is a tendency for the pressure to rise as the pump follows its flow/pressure curve. Relief valves will open to keep the pump flow above a certain level, so limiting the pressure they develop. Variable speed pumps will have a similar effect. As demand reduces, the speed of the pump can be slowed down to prevent excess pressure. It is also worth considering having valves on the outlet of pumps, which open only after the pump has started – again to prevent pressure surges.

Reservoir inlet control

Valves on the inlet to service reservoirs can be used to control the flow in relation to stock. They can be fitted with pilots or electronic controls to prevent pressure surges when they shut off as the reservoir fills.

Reservoir outlet controls

It may be possible to install control valves on the outlets of service reservoirs to make a small reduction in pressure (maybe only 2 or 3 m), which will have an effect over a large area.

Day/night districts

In urban areas such as city centres it may not be possible to keep valves shut permanently due to the complexity of the network and the demands placed upon it. However, one solution is to shut some valves permanently, and to fit electrically-controlled valves on other links, which are shut only at night. This allows the daytime demand to be supplied without excess head loss, and at night the flow passes though a single feed via a pressure-reducing valve in order to reduce leakage.

In all cases it should be remembered that the key issues to be considered are:

- the size of the area covered by the scheme
- the ratio of current pressure to proposed pressure

The current level of leakage from bursts, and whether this should be reduced before the true benefit from the scheme is assessed.

7.8.2   Pressure monitoring points

Pressures should be monitored at the inlet to the district, both upstream and downstream of the PRV where fitted, at the average zone point (AZP), and at the critical or target points.

The AZP is the point which represents the average pressure in the system. The average property height (APH) may be estimated based on property elevation. Table 7.3 shows the calculation of APH for a district of 5000 properties.

Table 7.3 Calculating the average property height (APH).

Ground level contour (m)	Ground level range (m)	Percentage of properties	Number of properties	Property-metres (pm)
35	32.5–37.5	20	1000	35 000
40	37.5–42.5	25	1250	50 000
45	42.5–47.5	10	500	22 500
50	47.5–52.5	40	2000	100 000
55	52.5–57.5	5	250	13 750
Total		100	5000	221 250
Average				44.25m

Data logging for AZP would be taken at the nearest available hydrant or other suitable fitting to the APH point, and corrected for the difference in the actual to the calculated elevation.

The critical points are those places which would first run dry if pressure is reduced. These are usually the high points in the area, those most remote from the inlet to the district which will incur the greatest head loss, e.g. those supplied though long runs of small diameter service pipe (e.g. remote farms), or those which feed critical customers. There may be several critical points in a district and all should be logged or surveyed on the initial pass. Analysis of the logged result will indicate which point is the most critical, and this is then used as the target point in the area, i.e. the point which is used to determine the maximum extent of pressure reduction.

7.9   OPERATION OF PRVs

When pressure management is used as a major element of the leakage reduction programme, then most of the pressure management schemes will require the installation of control valves. This section considers the issues associated with the selection, installation and maintenance of valves to control pressures.

7.9.1   Other uses of automatic control valves

Automatic control valves are employed as part of the leakage and network management strategy for several purposes other than pressure reduction. Gaining effective control of flows and pressures in the water distribution system, is an important element of an integrated strategy, which provides a more stable flow and pressure regime.

As well as PRVs, control valves are used to:

- Regulate the flows into service reservoirs and tanks. These valves can be controlled with simple hydraulic devices such as floats to shut off the inflow when the tank is full, or they can be controlled electronically to profile the inflow according to demand on the system and other parameters which are monitored continuously.
- Control the operation of pumps.
- Sustain pressures upstream of the valve in order to maintain a minimum level of service to customers or to prevent negative pressures, e.g. where the main runs over the top of a hill.
- Control flow rate e.g. on the supply to an industrial customer who takes water through a tank. The peak flow which occurs when the ball valve on the tank opens can cause pressure surges in the distribution system. This sudden increase in flow can be controlled by a valve which is set to prevent the flow exceeding a set limit.

7.9.2   Selecting and sizing PRVs

There are several issues to be taken into account when selecting the type of valve to be used for pressure management. Some water suppliers prefer to standardise on a single make of valve, whilst others prefer to specify a short list of suitable manufacturers. Single source has certain benefits:

- Training can be included with the valve supply contract. It is useful to spread the knowledge of PRV commissioning and maintenance around a number of staff, and for all staff working on routine distribution system operations to have an awareness of PRVs. So, limiting the number of valves in use helps to prevent a situation in which only a few staff have full knowledge of these technical issues
- Single source supply can give costs savings under competitive tendering
- Inter-changeability of parts and valves
- Reduced number of spare parts required
- Reduced maintenance costs

When choosing which valve(s) to use, the following factors should be included in the selection process:

- Reliability: does the valve have a proven track record of reliable operation?
- Quality: does the valve meet the necessary quality standards, whether these are set by the water supplier, or whether they are mandatory regulations which have to be complied with e.g. water quality standards?
- Suitability: is the valve suitable for the purpose?

- Life expectancy: does the manufacturer give any guarantees, or can he supply client references to indicate the expected life of the valve before it will require replacement?
- Maintenance: what are the maintenance requirements for the valve. These will have to be taken into account in assessing the cost–benefit analysis, and also the ongoing staff resource requirement?
- Standardisation: is there a single standard valve for each size, or are there different patterns and specifications depending on the installation site? Some valves have different springs for different outlet pressure ranges; some require additional or different inserts to cope with certain duties.
- Support: is support available from the manufacturer or his agent, to help with the design and commissioning?

Sizing of valves

PRV sizing is often a compromise between a valve which is large enough for the peak flow rate, and one which is small enough to give a stable outlet pressure at low flows. Low flow stability is a key issue with pressure-reducing valves. At low flows, it is more difficult to control the position of the valve to give the required head loss between the current upstream pipeline pressure and the set outlet pressure. In situations where the flow rate and inlet pressure is changing rapidly, this is more difficult. Instability can cause the valve to 'hunt' around the outlet pressure, it may lose control completely leading to major pressure transients, or it may vibrate leading to damage in the longer term.

Different manufacturers have different methods of controlling low flows. Some valves require the addition of a small diameter by-pass valve. The low flow goes through the small valve with the larger valve closed, until the flow rate reaches such a point that the main valve opens up. Some manufacturers have different pattern valves e.g. reduced port for low flow duties. Others use a factory-fitted throttling plug as a standard feature of the valve, or a slotted plug which is fitted only for low-flow duties.

Most manufacturers supply software to aid the selection of the valve size and model. The data which are required are:

- maximum recommended continuous flow (m^3/hr) which occurs at peak demand time, or due to industrial use
- minimum flow (m^3/hr), which usually occurs at night
- ratio of inlet pressure to outlet pressure. A large ratio may cause the valve to suffer cavitation damage (see section 7.12)
- acceptable head loss at peak flow. Where the differential pressure across the valve at peak demand is insufficient for a 2-way pilot diaphragm valve, a 3-way pilot may be needed.

- size of main. The valve will not be larger than the main, and due to the requirement for low flow stability it may have to be smaller. Some types of valve are capable of controlling low flows in such a way that they can be sized to the same diameter of the main, and therefore they give lower head losses at peak demand.
- size of meter. Often the meter is smaller than the main to which it is fitted in order to gain low flow accuracy. It is common practice for the PRV to be the same size as the meter.

7.9.3   Pilot arrangements

The valve used most commonly for pressure reduction in conjunction with district metering, is the diaphragm type. This type of valve creates a restriction to cause a head loss. The degree of head loss depends on the flow rate and the position of the rubber diaphragm. The simplest form of diaphragm valve is called a 'fixed ratio', in which the outlet pressure is a fixed proportion of the inlet pressure, regardless of flow rate. Changes to inlet pressure will be mirrored in the outlet pressure. The valve requires no pilot. The degree of head loss depends on the ratio of the diaphragm area to the area of the valve seat.

Fixed outlet valves use a pilot arrangement, which can be adjusted to give a set outlet pressure, regardless of flow rate or inlet pressure variations. A '2-way' pilot requires a continuous flow of water though a pilot loop between the upstream and downstream side of the valve. The chamber behind the diaphragm is always pressurised, and as such the valve cannot open fully. So, there will always be a head loss across the valve. In situations when the outlet setting of the valve is only marginally above the inlet pressure (usually at peak demand times when head losses in the distribution systems are at their greatest) then this head loss may be more than is acceptable. In these situations, a '3-way' pilot may be required. With this arrangement, the pilot allows the chamber behind the diaphragm to empty to atmosphere, and so the valve can open fully giving the minimum head loss.

7.9.4   Commissioning of valves

Based on experience in the UK the following procedure is recommended for commissioning pressure-reducing valves which control the supply to a given area.

1   Install data loggers in the area to monitor pressures during the commissioning stage. Ideally, these should be left in place until commissioning is complete, and can be downloaded after each change to the pressure regime to determine that adequate supplies are being maintained. If such equipment is not available, then it will be necessary to undertake pressure checks manually using a gauge perhaps connected to a hydrant.

2   The valve will usually be installed on a by-pass downstream of the district meter. The valve should be commissioned fully open i.e. the water should flow through the valve, but the pilot is set so that there is no reduction, or one of the cocks on the pilot pipe work is set to the fully open position.

3   If there is a general programme of leak detection and repair in operation, then the area should be surveyed to find and fix leaks to bring the leakage level down to the set cost 'exit' level.

4   Immediately after the detection and repair works are complete, the PRV should be set to make the first reduction in pressure. The size of the first reduction will depend on the difference between inlet and outlet pressure. Some practitioners prefer to make a set number of reductions, in which the pressure drop is achieved in equal stages (between 2 and 5 stages is common practice). Others prefer to make the first reduction as a set percentage of the full reduction (say 50%). Subsequent reductions are again a fixed percentage of the remainder (again 50% of the remaining drop) and so on until there is only a small final reduction to achieve the desired outlet setting.

5   During the reduction stages, customer calls should be monitored and corrective action taken if pressures in part of the area, or to individual customers fall below acceptable levels.

6   Once the required outlet setting has been reached, the area should continue to be monitored for a period of a few days to ensure there are no outstanding issues to resolve

If the area is to be created specifically for pressure management, then the first stage may be to field test the new area before the valve is installed. The area will be created by closing boundary valves. Data loggers placed at the key points in the area will show whether all customers are receiving acceptable pressure, and they will indicate the scope for pressure reduction.

7.9.5   Reduction in pressure

Other than critical customers, a typical schedule of acceptable pressures may be as follows:

- Aim for 30 m at the target point at peak time on most days.
- Accept 25 m on some days.
- Dips to 20 m may be noticed at the critical points in extreme situations.
- Ensure that no property receives less than 15 m at any time.
- Pressure should not be reduced in a single step, otherwise there could be too much of an adverse effect on customer supplies: the recommended approach is described in section 7.9.4.

- A period of 7 days should be left between each reduction in pressure to allow for the effect to be monitored and any issues to be dealt with. Where large reductions are made, the first steps may be made within 3 or 4 days of each other if no problems are experienced.
- The district should be monitored by data logging or pressure survey while the pressure reduction is being carried out. Ideally, data loggers should be left in place at the key points in the district, and these should be downloaded at intervals to determine whether the effect is as planned.
- Customer complaints should be monitored, and dealt with after each stage.

7.9.6   Potential problems with valves

Modern diaphragm pressure-reducing valves are reliable pieces of equipment. However, problems can arise typically:

- valves can fail open if there is a blockage on a failure on the inlet side of the pilot pipework
- valves can fail shut if there is a failure on the outlet side of the pilot pipework
- valves may become unstable due to malfunction of the pilot
- the diaphragm may split due to wear and tear or due to damage caused by debris carried in the water main

7.9.7   Monitoring and maintenance of valves

All pressure-reducing valves will eventually require some form of maintenance. Unlike sluice valves which are often installed and left alone, PRVs are small machines which run continuously. Improvements in the design of PRVs has resulted in units which need relatively low maintenance. In order to minimise the risk of valve failure or malfunction, most manufactures recommend a programme of periodic inspection and maintenance. The extent and frequency of the maintenance required varies quite significantly from one manufacturer to another, and therefore it is a key issue when selecting which valve to use.

Typical maintenance includes:

- inspecting and cleaning the filter which protects the pilot from debris
- changing 'o' ring seals
- changing the diaphragm
- inspecting and cleaning the pilot valve

Maintenance intervals vary typically from 6 months to 2 years. The choice for the water supplier is between the costs of maintenance and the risk and consequence of failure of the valve. An appropriate strategy is to categorise valves into high and low risk groups, and to carry out more frequent maintenance on those valves which are more likely to fail, or for which the impact of failure on customer supplies is a major issue.

Even with a programme of maintenance, valves can fail due to some form of damage or extreme event. Therefore, it is recommended that the operation of all valves be monitored at regular intervals. Monitoring may take the form of regular visits to the valve to check the outlet gauge setting, logging the valve at intervals, continuous logging, or the installation of telemetry to provide the data on-line. Again the choice will depend on cost and risk.

7.10   FLOW MODULATION

Where valves are installed with a fixed outlet, head loss in the area under control means that the pressures within the network will vary with customer demand. The outlet setting is governed by the pressure required at the critical point(s), such that at peak demand time, when head loss is highest, the minimum pressure at the critical point is still within acceptable limits.

This arrangement means that when demand is less than peak, the pressures in the network will be above the requirements to meet standards of customer service. Flow modulation is a method of reducing the excess pressures, and by doing so making further reductions in leakage.

Flow modulation systems can be used to fix the pressure at the critical point, by adjusting the outlet setting at the inlet valve to compensate for head loss. By implication, this gives a fluctuating pressure at the inlet to the area, and it should be recognised that it is not possible to keep pressure fixed at all points in the network. Therefore, a compromise has to be made between the inlet pressure and the critical point pressure in order to optimise the leakage level.

Figure 7.9 shows the pressure profile at a critical point in the system with a conventional fixed outlet PRV, and with a flow modulated PRV. The flow modulated PRV provides a more constant pressure, which is less than the pressure applied by the fixed outlet valve at most times of the day. At peak demand time, the pressure exceeds that of the fixed outlet valve. Flow modulation allows average pressures to be reduced below those from an equivalent fixed outlet valve, allowing greater reductions in leakage to be achieved.

There are many different types of flow modulation device, some of which are outlined below.

Electronic remote control

A device at the critical point monitors the pressure and sends a signal back to a control unit at the PRV. The relationship between critical point pressure and PRV

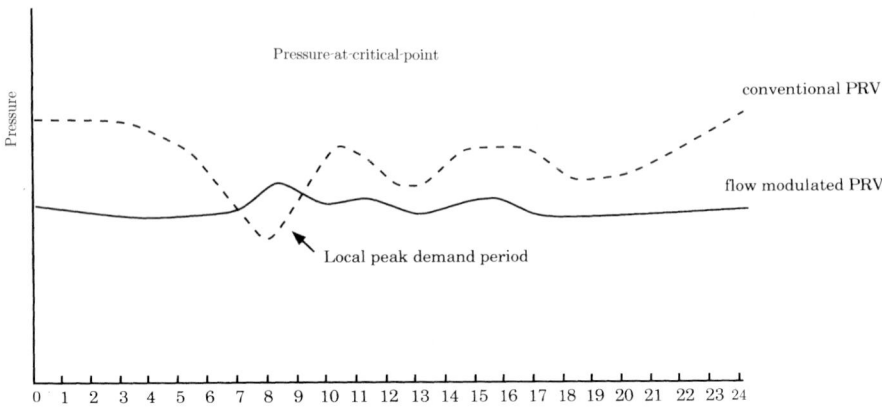

Figure 7.9 The flow modulation concept,

setting can be programmed into the controller. The signal can be transmitted by hard wire, or by radio or mobile telephone systems.

Electronic local control

The local controller takes a signal from the flow meter, and the relationship between flow rate and outlet setting is programmed in to the control unit. If no flow meter is available, the unit can be programmed to adjust the PRV setting for different times of the day, based on a typical flow profile. However, with this arrangement, a sudden change of flow e.g. a demand for water for fire fighting, could result in low pressures in the area as the PRV controller would not respond to the new situation.

Hydraulic devices

Hydraulic devices can be used to give continual changes to outlet setting without the need for a flow meter or electronics. They work by sensing a change in head loss across an orifice plate, or through a Venturi tube as flow rate increases. This change in head loss is used hydraulically to change the outlet of the PRV pilot.

Continuous adjustment

Some flow modulation devices can be set to adjust the PRV setting 24 hours a day. These are either connected to a flow meter signal to adjust the valve in accordance with the actual demand on the network or they are pre-programmed to give a set outlet at a set time of day. The settings are programmed into the modulator, based on analysis of the flow and pressure regime in the district on a typical day. Adjustments can be made to allow for high demands on unusual days such as bank holidays.

Day / Night

Simple controllers make only two adjustments a day to the pressure setting, usually a day setting and a night setting. These controllers are generally less expensive, but they give a benefit of reducing night pressures when leakage is at its highest.

7.11   FACTORS RESTRICTING THE SCOPE FOR PRESSURE REDUCTION

When designing a pressure management scheme there are usually a number of critical factors which limit the scope for pressure reduction. These have to be managed well or the potential benefits may not be obtained, or there may be problems with customer supplies.

Tall buildings

The area should be surveyed to identify tall building such as high rise apartments, office blocks, and industrial premises. The method by which these are supplied should be determined. If there is a basement tank from which water is pumped up the building, there will be little impact from the pressure management scheme. However, if the supply relies on the mains pressure to feed the top floors, or pumps part way up the building, then the extent of pressure management could be severely limited.

Tall buildings can be identified from mains records or billing records initially, and these should then be confirmed by a visit to the area.

It may be cost-effective to change the internal plumbing layout in buildings if this will allow a greater scope for pressure reduction and hence a saving in leakage.

Even if the scheme can go ahead with the existing pipework layout, there is a possibility that the booster pumps will operate more frequently or they will operate at a lower inlet pressure. This will create a higher cost to the property owner, who should be informed in advance.

Critical industrial customers

Large water users should be consulted to evaluate their water demand, the times at which they take peak flows, and the consequent effect of any pressure reduction programme. Working with customers may allow schemes of mutual benefit to be developed, whereas proceeding without consultation may lead to conflict resulting in delays or abandonment of proposals.

Fire supplies generally

Consultation will have to be made with the relevant fire fighting authorities, and if done correctly then it is unlikely to cause a major impact on pressure management proposals. The important message to get across is that flow capacity is far more important than static pressure. High pressure on an inadequately sized main will be far less effective for fighting a fire than a hydrant connected to a large main running at a lower pressure.

Some fire authorities use water direct from hydrants, whereas others use the hydrant water to recharge tenders and tankers. If there are guaranteed standards then the water supplier must either accept that they will restrict the scope for pressure reduction, or negotiate changes with the fire authority.

Correct sizing of the PRV is critical if the effect on fire flows is to be minimised, and real time flow modulation may also be used to supplement falling pressures caused by the increased demand on the network.

The water supply organisation may reach an agreement with the fire authority setting out procedures to be followed in the event of a fire e.g. by-passing the pressure-reducing valve.

Specific sprinkler requirements

Some industrial premises are fitted with fire fighting sprinkler systems, and there may be written standards for these set by the insurance company. Sprinkler systems may need adjustment to allow for the revised pressure regime.

Health authority and special needs customers

Health authorities should be consulted to assess the impact on hospitals, clinics and home dialysis patients in the area supplied, as these supplies are critical and may reduce the potential for pressure management .

It is important for the water supplier to develop well-considered and appropriate policies, and guidelines for staff, before implementing pressure management, in order to ensure a consistent approach and to avoid conflict with customers.

7.12   CAVITATION

Cavitation occurs when the pressure of water falls below its vapour pressure at a particular temperature. At this point the water effectively boils and small bubbles of vapour form in large numbers. These bubbles are carried in the flow and when they reach a point where the pressure recovers to a higher level, they suddenly collapse as the vapour condenses to liquid again. A cavity results and the surrounding water rushes to fill it. The water moving from all directions collides in the cavity, giving rise to very high localised pressures. Any solid surface in the vicinity of this effect is subject to these very high pressures. The alternate formation of cavities and high pressures can occur at high frequencies of thousands of times a second. The impact can cause severe damage to metal internal surfaces of valves and pipes, when the material fails due to fatigue after a long period of exposure, aided by corrosion, and wear due to the high velocity of water. The metal becomes pitted and even larger pieces can be torn away. The cavitation is often accompanied by severe noise and vibration.

Put simply, cavitation can occur to some extent in most valves, but it has to be limited in order to avoid reducing the life of the valve. There are two fundamental ways of managing the problems:

1   To design out the effect by managing the parameters which lead to cavitation.
2   To design valves in such a way as to minimise the effect of cavitation. Use of resistant materials such as stainless steel for the key components will help avoid cavitation damage. The shape of the valve also has an effect, and the impact will be reduced if there is sufficient space within the valve body for the cavitation to occur without it impacting on the internal wall. This is one benefit of being able to install a valve the same size as the main to which it is attached.

The cavitation index is calculated by the following formula:

$$K = \frac{H_{inlet} - H_{outlet}}{H_{inlet} * 10}$$

where K = the cavitation index
               H_{inlet} = the inlet pressure (metres head)
               H_{outlet} = the outlet pressure (metres head)

When K is above 0.3, the noise created due to cavitation will increase to noticeable levels. When the index is above 0.6, severe cavitation damage can occur.

The index is governed mainly by the ratio of inlet pressure to outlet pressure. When this ratio is above 3, further consideration should be given to the issue. When it is below 3, the practitioner should not be unduly concerned.

There are two solutions:

1   To fit two valves in series in order to reduce the head loss created by each valve. The first valve can be a simple ratio valve, with the second valve being used to give the required outlet setting.
2   To fit an orifice plate to the outlet of the valve to create a back pressure. The plate will only be effective above a designed flow rate, required to create sufficient head loss. Care must be taken to ensure that cavitation damage does not occur to the plate itself.

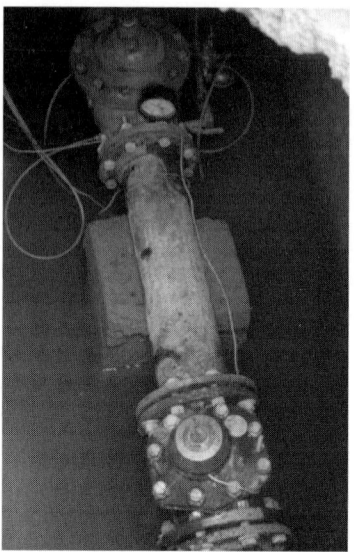

Figure 7.10 PRV and meter installation.

7.13   PRV INSTALLATION

PRVs are usually sited next to the district meter. The PRV should be downstream of the meter so that the turbulence from the valve does not affect the accuracy of the meter. It is good practice to install the PRV on a by-pass to enable future major maintenance to be carried out without reversing flows in the district. It is common practice to put the PRV on the by-pass pipework to allow the chamber to be built in a suitable location with convenient access. Whilst the main may run in the road carriageway, the chamber can be built in the footpath or verge. There are a number of arrangements which are often used, as shown in Figures 7.10 and 7.11

The site of the PRV could also be remote to maximise valve performance and to ensure that customers receive an adequate pressure. For example, if the main into a district runs downhill, the PRV may be installed part way down the hill so that higher properties continue to receive full pressure. Boundary valves have to be shut inside the DMA to create a pressure-managed area (PMA).

Logging points should be provided for monitoring the PRV operation. These may be fitted to the valve itself, but if not they should be provided either inside the chamber or on branches either side of it.

Once a suitable location has been found, a construction package should be prepared for the installation contractor containing a plan of the zone, a

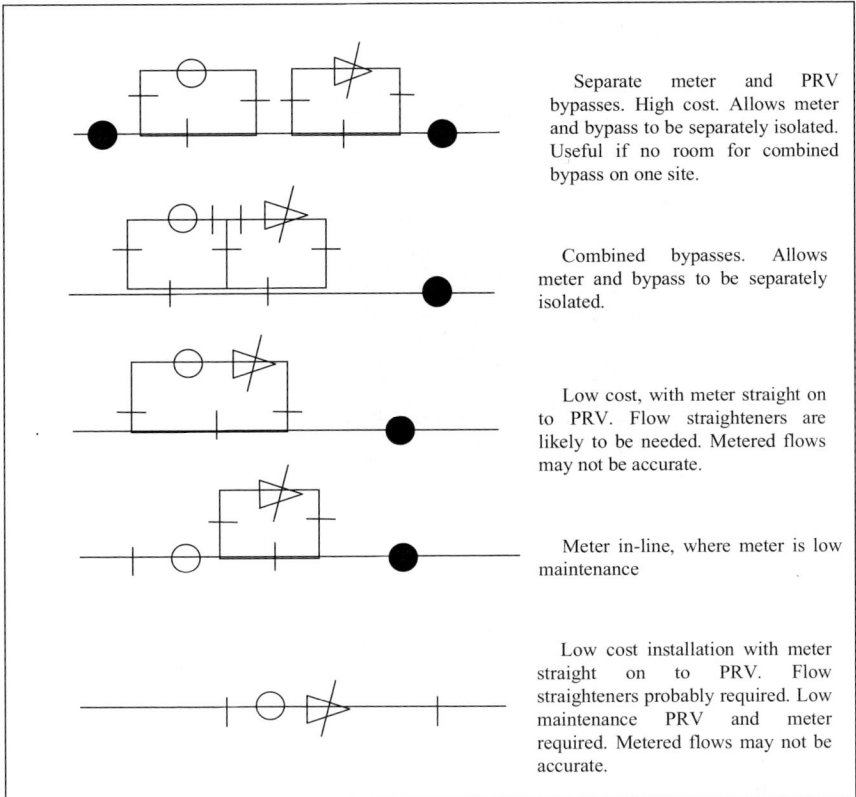

Separate meter and PRV bypasses. High cost. Allows meter and bypass to be separately isolated. Useful if no room for combined bypass on one site.

Combined bypasses. Allows meter and bypass to be separately isolated.

Low cost, with meter straight on to PRV. Flow straighteners are likely to be needed. Metered flows may not be accurate.

Meter in-line, where meter is low maintenance

Low cost installation with meter straight on to PRV. Flow straighteners probably required. Low maintenance PRV and meter required. Metered flows may not be accurate.

Figure 7.11 A selection of meter and PRV installation layouts.

detailed drawing of the installation, and any other relevant information such as utility plans.

7.13.1 Potential problems

There are a number of problems which should be anticipated when implementing a pressure management scheme:

- Customer complaints of low pressures or no water can occur even though the pressure in the mains is adequate. This may be due to a problem with the service pipe such as
  - a blockage at the stop tap or meter

- an inadequate size or damaged service pipe
- a leak. Stop taps should be checked to ensure they are fully open and that there is no leak in the service connection.
- Unexpectedly low average zone (AZ)/target point (TP) pressures may occur due to unknown shut valves within the network. All line valves should be checked to ensure they are open
- Passing boundary valves will result in water being drawn in from adjacent districts when pressure in reduced. A test (sometimes called a pressure zero test – PZT) should be carried out to ensure the district is tight before pressure reduction is implemented. The inlet(s) to the district is closed (usually at night), and the pressure is monitored to ensure that it drops to zero. Failure to do so, may indicate a passing boundary valve, or an unknown connection.
- Erratic meter/PRV performance can result from debris collecting in the mechanical parts after it has been dislodged from the mains pipework due to the flow reversals caused by the valve operations in the network.
- Pressures may be lower than expected because of an unknown existing pressure control valve, or may be higher than expected because of an unidentified booster pump.
- Poor infrastructure condition or low capacity may cause other properties to be affected

A plumber should be available to resolve problems inside customers' premises; perhaps a boiler or appliance which does not work at the lower pressure, or a blocked ball valve or stop tap. Domestic appliance pressures vary from country to country. Typical values are as follows:

- float valves work satisfactorily up to 14 bar
- dishwashers/washing machines 0.5 bar to 10 bar
- electric showers 0.7 bar to 10 bar, but poor below 2 bar
- un-vented hot water systems (e.g. combination boilers) optimum at 2.5 to 3 bar and minimum flow of 20 l/min

7.13.2 Post-project audit

On completion of a pressure management scheme, a post-project report should be prepared setting out all the relevant data. This is best carried out scheme by scheme, in line with the design packages. The report should contain information relating to:

- the size and type of PRV installed
- the actual boundaries of the area that has been pressure reduced, together with the property counts and other statistics
- the before and after logging data showing pressures, flows and calculated leakage levels

- the setting of the PRV
- an assessment of the leakage savings obtained from the scheme

7.14   PREDICTING A PRESSURE-MANAGED REGIME

Several software packages are available to predict the impact, savings (volumetric and financial) and the payback period set against the capital cost of installing a PRV or other type of pressure management scheme. The data required will vary from package to package, but the common data are generally:

- pressures upstream and downstream of the PRV
- flow through the meter by logging or estimate (logging at intervals of 15 minutes is common, but a frequency of 5 minutes or less gives a better indication of any instability or sudden changes)
- minimum 7 days data – avoid seasonal effects
- demand profile and any potential changes
- schedule for stepped reduction in pressure
- inlet, AZP, target point pressures
- standard DMA data such as distribution mains length, number of household properties, and the number of non-household properties

Predictive software operates most efficiently at DMA level, but it can also work at a supply zone level. The principles can also be incorporated into the software used to set zonal targets based on economic principles.

The data relating to the current supply regime is input, and then the option is given to select the future pressure management regime, either with the introduction of a fixed outlet valve, or a flow-modulated one, or by making changes in other ways. The proposed pressure profiles are input, from which the predicted savings in leakage are calculated. Packages vary in degrees of sophistication. The impact on pressures at the AZP and critical points can be seen from the profiles. Some packages will also give supplementary data used in the leakage analysis:

- verification of the per capita consumption estimate based on a comparison of night flows with average daily flows
- the hour to day correction factor (existing and future) as described in section 7.15
- an estimate of the excess leakage expressed in Ml/d, litres/property/day, or as ESPBs (see Chapter 3)
- an estimate of the payback period for the scheme based on input data from the value of water, and the costs of the scheme

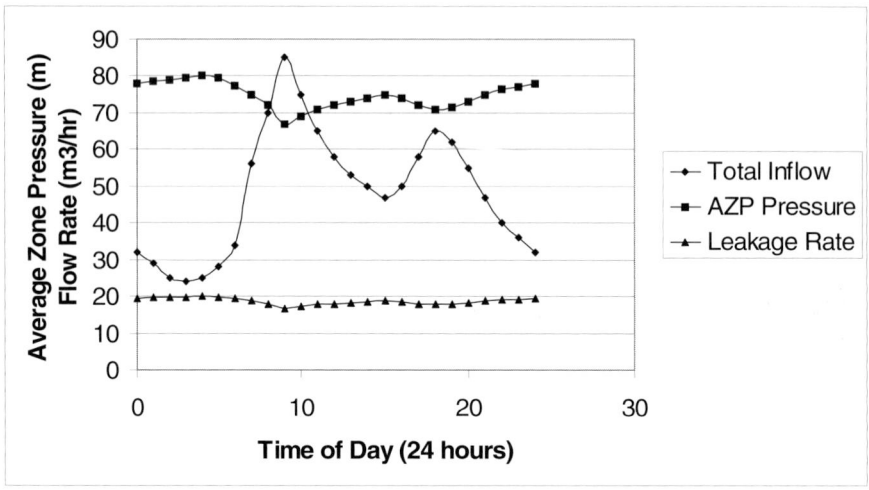

Figure 7.12 Variation in AZP in a gravity-fed system – hour to day factor = 22.5.

7.15   HOUR TO DAY FACTORS

When leakage estimates are based on nightflow measurements, the value measured in m³/hour has to be converted into an average volume for the day, which can then be compared to the value from the annual water balance calculation. Simply multiplying by 24 hours can give an over or underestimate of daily leakage, due to differences in pressure. At night when the leakage estimate is made, the pressure tends to be at its highest in gravity-fed systems, when head loss due to demand is at its minimum. Therefore the leakage rate at night, when the estimate is made, will be higher than the average over the day. To take account of this fact, leakage estimates based on nightflow measurements are multiplied by a factor, usually known as the 'hour to day' factor.

For gravity-fed systems the factor is usually less than 24. STC Report 26 [6] recommended the use of a 20-hour day, based on a limited sample of data. More recent research suggests that in the absence of any detailed understanding of pressure variations in the supply zone, a 22 hour to day factor is more appropriate. A factor of 22 is often used for calculations for large supply zones or for the whole of an organisation, until such time that detailed data becomes available from site measurements.

In some systems, the pressure during the day is higher than at night e.g.:

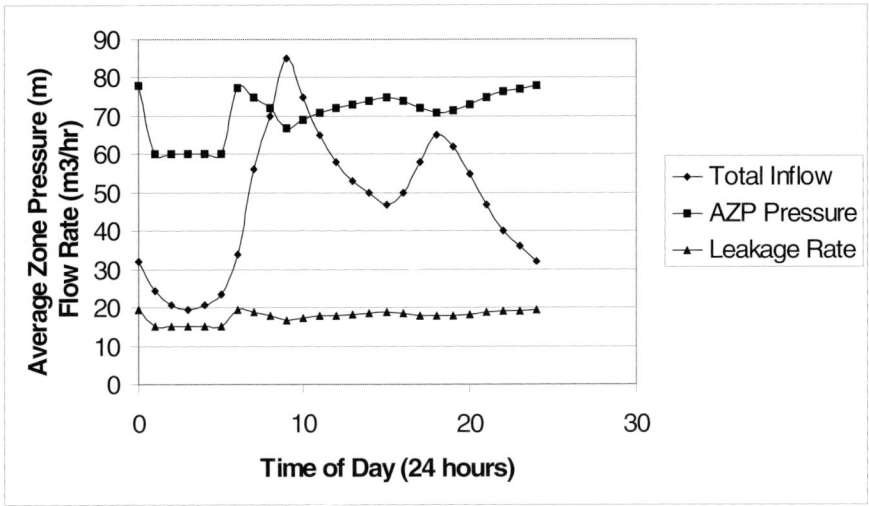

Figure 7.13 Variation in AZP in a pump-controlled system – hour to day factor = 29.5.

- when a flow modulated, or time modulated pressure reducing valve has a lower pressure setting at night, than during the day when a higher setting is required to overcome head losses in the network
- when a pump is used to supply daytime demand, and which can be switched off at night.
- when control valves are opened by telemetry, or sometimes manually, to meet daytime peaks, or to control the flows between service reservoirs.

In these cases, the hour to day factor will be more than 24.

In the UK, it is common for systems to have districts with hour to day factors ranging from 15 to 30, with the average being close to 22. This variation has been noted in other parts of the world.

Figures 7.12 and 7.13 show the relationship between the inflow to the district and the pressure at the AZP point for a gravity-fed system and a pumped system. They also show how the leakage rate is influenced by the pressure.

7.16   REFERENCES

1   Lambert, A (1997) 'Pressure Management/Leakage Relationships: Theory, Concepts and Practical Application'. Paper presented at IQPC Seminar, London, April.

2   Goodwin, SJ (1980) *The Results of the Experimental Programme on Leakage and Leakage Control,* Technical Report TR154. Swindon: Water Research Centre.
3   May, J (1994) 'Leakage, Pressure and Control'. Paper presented at BICS International Conference on Leakage Control Investigation in Underground Assets, London, March.
4   Lambert, A (2001) 'What do we know about pressure: leakage relationships in distribution systems?' *Proceedings of IWA Conference on System Approach to Leakage Control and Water Distribution Systems Management*, 16–18 May 2001, Brno: Brno University of Technology.
5   WSA/WCA Engineering and Operations Committee (1994) *Managing Leakage: UK Water Industry Managing Leakage* Reports A–J: Report A – *Summary Report*; Report B – *Reporting Comparative Leakage Performance*; Report C – *Setting Economic Leakage Targets*; Report D - *Estimating Unmeasured Water Delivered*; Report E – *Interpreting Measured Night Flows*; Report F – *Using Night Flow Data*; Report G – *Managing Water Pressure*; Report H – *Dealing With Customers' Leakage*; Report J – *Techniques, Technology and Training.*London: WRc/WSA/WCA.
6   Technical Group on Waste of Water (1985 [1980]) *Leakage Control Policy and Practice*, Standing Technical Committee Report no. 26. Original publication London: Doe/NWC. Reprinted London WAA/WRc.

8

Changing policies

At some stage, in all organisations, it becomes necessary to examine the policies for producing and delivering water. Some policies relate to managing elements of the infrastructure – pipework characteristics and condition, and the way in which it is operated and maintained – upgrading and managing the infrastructure has been addressed in previous chapters.

Other policies are largely organisational – they relate to how the company views its relationship with its customers, and having the appropriate staffing and regulatory frameworks in place to deal with its main function – to produce and deliver water to its customers. Such policies are very subjective – they are influenced not only by the physical and local characteristics of the network, and the social and cultural attitudes of the customers, but by the structure of the company itself, whether public or privately owned, or public/private sector partnerships. In this case the organisation will have other drivers to consider, such as the interests of directors, shareholders, political and financial pressures, as well as customer and public perception. There are also increasing environmental risks of balancing new resources against the need to meet ever-increasing customer demand. Such policies include:

- demand management and water conservation
- regulatory and legal frameworks
- customer metering policy, tariff structures, and revenue collection

These issues are dealt with in the following sections, and are illustrated in the case study examples.

8.1   CUSTOMER DEMAND

Figures provided by the UN [1] show that the world's population trebled in the twentieth century, leading to a six-fold increase in the use of water resources. In global terms the three largest water users are:

* agriculture – 67%
* industry – 19%
* municipal/residential – 9%

As the amount of disposable income increases so generally does the amount of water we use. For example, the large majority of households in the developed world now have washing machines; the ownership of dishwashers is increasing; the huge increase in car ownership means more water is used for car washing and a growing interest in our external domestic environment means that more water is used for garden watering.

But it is not only affluence that is increasing our demands for water. In many countries the population is rapidly increasing. Already 1.1 billion people have no access to clean drinking water [1] Populations are also migrating. The UN is currently forecasting that more than 60% of the world's population will be living in urban areas by 2030. In the less-developed regions of the world urban populations are set to more than double. Providing a clean supply of drinking water to these migrants alone will place a considerable demand on what, in many places is an already over-stretched supply. Failure to do so will result, among other things, in the rapid spread of disease.

In the same way that the per capita consumption varies between different countries because of economic, social, religious, geographical and a number of other factors, so to does the way in which water is used. Even within a small area there will be considerable variations between the use of water in different households dependent, among other things, on the number of occupants, socio-economic class, and whether or not the water supply to the property is metered. It is even more difficult to quantify the proportions of water used for different purposes in industry, commerce and the service sector.

Figure 8.1 shows the ways in which water is used in domestic properties in the United Kingdom based on a survey of water use carried out by one water company in 1995 [2].

It is even more difficult to quantify the proportions of water used for different purposes in industry, commerce and the service sector. The consultation paper 'Using

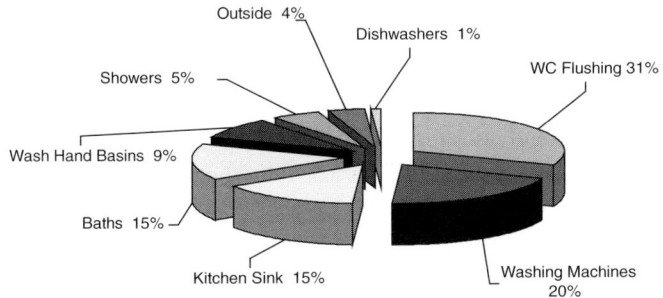

Figure 8.1 Components of household water use in the UK.

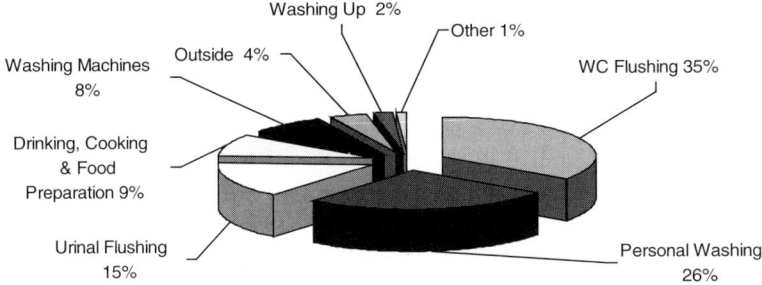

Figure 8.2 Components of industrial water use in the UK.

Water Wisely' made a number of assumptions, based on research by the Building Research Establishment (BRE) in 1978 [3]. This is shown in Figure 8.2. Although there may be quite substantial variations between individual units, these data give a good indication of where the most water is used and therefore where potentially the major savings may be made.

8.2   OPTIONS FOR MEETING INCREASING DEMAND

Apart from providing a suitable water supply infrastructure, in many areas it will be necessary to increase supply by tapping into new sources. Huge capital expenditure programmes to construct new dams and reservoirs or sink new boreholes will be required.

Major works such as these, as well as placing a financial burden on governments, municipalities and consumers, also carry with them environmental and social costs.

But in some areas the available natural resource may already have been depleted to the extent that further abstraction is either not possible or will be of such poor quality that it is extremely costly to treat. And in the future we may not be guaranteed the levels of precipitation that we have relied on in previous years. Climate change is occurring, but it is a largely unpredictable factor. Scientists believe that worldwide temperatures will rise between 1.0° and 3.5°C by the end of this century. Weather patterns are forecast to change with an increasing frequency of heatwaves, floods, droughts and storms in different parts of the world. Sea levels generally are also predicted to rise by between 15 and 95 centimetres. All these factors will have a major effect on the hydrological cycle.

There are of course other options available to increase supply:

- Water can be transported from areas of plenty to areas of need. It can be transported by pipeline or in containers, overland or by sea.
- Desalination is a further option in countries that are near to the sea.
- For some areas it may be necessary to set aside cultural and religious objections to recycling and treat sewage effluent to high enough standards to produce water that is fit for drinking.

But these options are not without high financial and environmental costs (and therefore higher costs to consumers or taxpayers). It is not cheap to transport water. Not only are long pipelines expensive to install, but water generally requires pumping and that carries a high economic and environmental cost because of the amount of energy required. Huge advances in membrane technology and waste heat recycling may have reduced the costs of early high-energy-use desalination plants, but this is still a relatively expensive way of providing wholesome water. Sewage treatment to standards that meet environmental requirements is one thing, but to bring that effluent up to wholesome standards without a considerable amount of initial dilution, as occurs along most major rivers, again requires expensive plant.

If new sources, transportation and recycling are impossible or highly expensive, then there is a need to look elsewhere for sufficient water to meet the increased requirements. Options include not only alternative low-cost sources, such as rainwater harvesting and shallow wells, but community education programmes to reduce waste and undue consumption of water, and to ensure that it is used in a sustainable manner. Reducing the demand for water also has the added benefit of reducing the volume of sewage that has to be treated. It could relieve problems at sewage works that are working to capacity and postpone costly extensions or new sewage treatment works.

8.3   DEMAND MANAGEMENT POLICIES

In commerce, industry and agriculture, and also in metered domestic premises, reducing water usage without compromising standards can result in considerable reductions in the cost of water supplied. In addition, where wastewater charges are based on the volume of water supplied, a double saving will be achieved through a reduction in wastewater charges. Although the cost of water is generally small in comparison with the turnover of an organisation (about 1–2%), a simple demand management programme can typically achieve savings of 20–50%. In some cases savings of up to 80% have been recorded.

To be successful, a demand management programme does not have to impose restrictions on all customers. Neither should it compromise health and hygiene. But it does require *recognition* by the water supplier and its customers that water is a scarce resource, and *efforts* on both sides to reduce waste, misuse and undue consumption. This includes:

- a user education and awareness programme
- the development of efficient systems to manage water networks
- ensuring there are no illegal connections
- ensuring all supplies are metered and that charges for water reflect actual costs of supply
- retrofitting water efficient appliances and fittings
- introducing recycling systems
- setting legal standards for water using appliances

Implementing a demand management programme in a water supply authority, a public body or industrial, commercial or agricultural organisation, requires a high level of commitment from top management. It also requires a dedicated team with a highly motivated leader. The benefits are measurable and realistic targets can be set. Regular monitoring of performance against targets provides the necessary motivation.

8.3.1   Audit of water use

The starting point for any demand management programme must be a thorough *audit* of customers' water use, both for domestic and commercial users. Although a great deal of consumption data is available for water use in various situations and for many appliances, as shown in Figures 8.1 and 8.2, it is much more difficult to assess the components of commercial and industrial use. These are highly dependent on the nature of the business, the number of employees, and the process carried out by the business. Depending on the nature of the organisation and scale of the project an audit might require the installation of additional local meters and flow measurement devices. Guidance notes for conducting a water use audit are contained in Appendix D.

Once the audit of use has been completed a water balance can be calculated. This includes identification of all sources of water and the volume delivered from each source. Assuming that the supply and consumption data are accurate, any non-revenue water will be from leakage or, depending on the situation, possible illegal or unknown connections.

At the end of the audit, decision makers will be made aware of any anomalies in site consumption, wasteful practice, and leakage. They will be enabled to address the anomalies and to introduce best practice solutions.

Once usage has been established it is possible to prioritise the areas where the most savings can be made. In each case the cost of delivering the reduction in use and any additional maintenance costs must be viewed against the potential saving in the cost of water supplied. Some individual measures will have very short payback periods. The installation of a low-cost toilet cistern displacement device for example will have an almost immediate payback, while urinal control systems might have a 2–3 month payback period. Larger schemes, involving retrofit of low flow plumbing fittings (toilet cisterns, taps and showers) and recycling systems, might have a longer payback period of 1–2 years. Each case must be judged on its merits.

8.3.3   Public attitudes and user education

One of the major barriers to the introduction of any demand management programme is the attitude that people have towards the use of water. Because the raw material falls freely from the sky and naturally flows into our rivers and percolates into underground aquifers, too often consumers regard it as a plentiful and free resource. It is therefore used without constraint or consideration of the wider social and environmental repercussions of continuing to feed an ever-increasing demand.

User education therefore plays a key role in any demand management programme. It is important that users understand the hydrological cycle and are aware of the concepts of sustainability and why it is necessary to reduce our use of water. Achieving the necessary level of understanding requires different approaches for different types of organisation. It may require in-house seminars, literature drops, informal presentations and discussions at group meetings, television, radio and press advertising and integration of demand management issues into the school curriculum.

Education can also help to reduce usage by providing useful hints on the way we use water in situations where there is user control. For example, this might include such measures as:

- not leaving the tap running while cleaning teeth, shaving and washing
- not washing-up under a running tap
- not filling the kettle with more water than is needed
- only using the washing machine and dishwasher for full loads or where possible using half load/economy settings (saving energy as well).

It is also important that users have some idea of the quantity of water that is used in different operations or can be wasted through leaks and poorly maintained fittings.

8.3.4   Plumbing design

While considerable volumes of water can be saved by careful use an even greater impact can be made by installing water efficient fittings and ensuring that the plumbing and heating system is designed with water and energy efficiency in mind. Ideally this should be the case in all new buildings. Careful planning can save both water and energy. Regrettably many new buildings do not even conform to basic standards of water system design and incorporate low-cost fittings that lead to waste or undue consumption of water. Common inefficiencies include incorrect pipe sizing, long pipe runs from water heaters to taps, male toilets which have WCs and no urinals, poor-quality WC flushing devices that leak, cisterns and flushing devices that deliver more water than is necessary to clear the WC pan and convey waste to the sewer.

Good design will incorporate a range of water efficient fittings. These include:

- dual-flush WCs
- urinal flushing controls
- low flow (spray, sensor or push) taps in washrooms
- overflow pipes from cisterns to show discharge conspicuously and safely
- water-efficient appliances such as washing machines and dishwashers
- recycling of rainwater and greywater for WC flushing and garden watering
- flow regulators and devices to detect abnormal water flows

8.3.5   Retrofitting water efficient devices

But it is not only in new buildings that efficiencies can be achieved from systems and fittings. There are several examples where large scale retrofitting with efficient water fittings has made considerable reductions in water demand. One such initiative was New York, in 1997. A major programme of changing the 18-litre toilet cisterns to 6-litre ones, and fitting low-flow showerheads in apartments, has resulted in reductions in these dwellings of some 30% of water usage. Most water companies in the United Kingdom have provided water displacement devices to their customers for fitting in existing toilet cisterns. In most cases these have been successful in reducing demand. However, it is essential that the device is fitted correctly without interfering with the flush or water inlet mechanism and that if double flushing becomes necessary the device is removed immediately.

Considerable savings can be achieved in existing non-domestic premises by fitting urinal controls. These include:

- infra-red devices on individual units which activate a low-volume flush each time the urinal is used
- programmable devices that activate a flush periodically but only during the time that the premises are occupied

Waterless urinals might also be considered, but due consideration must be given to the economics and the fact that they require cleaning and in some cases replacement of odour-destroying pads, parts or chemicals. If the additional maintenance cost outweighs the saving in water costs then some other more cost-effective solution should be used.

Fittings that restrict the flow of water are readily available and, where their use is appropriate, can quite easily be fitted to many appliances. Examples might include where articles are washed under a running tap or perhaps where taps might be left running by careless or inconsiderate users (although self-closing or sensor-controlled taps might be better in such cases).

There is a considerable debate about the effectiveness of installing showers as opposed to baths in domestic properties. On average baths require about 80 litres of water and a conventional shower might use about 30 litres. On the face of it showers are more economic in both water and energy use. However, there is a tendency for households with showers to use them more frequently than households which only have baths. And, there may be a tendency to stand under a hot shower for a considerable time. Often the main control over how much water is used in a shower is the volume of hot water available. There is also a tendency to install pumped showers, or showers with multiple showerheads, which are not as efficient in water use as conventional showers.

Low-flow showerheads can reduce the volume of water discharged by a conventional showerhead by about 50%, depending on the supply pressure. Installing a low-flow showerhead is likely to be a better option than installing a flow restrictor because the low-flow showerhead is specifically designed to give an adequate drench of water whereas the flow restrictor may result in an inadequate flow and therefore increase the showering time.

Considerable volumes of water can be lost from leaks and overflows during periods where premises are left unoccupied, such as over a weekend or during a school holiday. Premises that are unattended and unheated during the winter months are particularly liable to suffer bursts in water pipes during periods of prolonged freezing. Water losses of this type can be minimised by the installation of a device that can be programmed to shut off the water supply where an abnormal flow is detected.

8.3.6 Legislation, regulation and enforcement

Despite the benefits that can be achieved from a demand management programme, there is often considerable reluctance on the part of water suppliers and customers to implement them. Where water delivery is the sole responsibility of the private sector, there are clearly commercial advantages in selling as much water as possible. But, if the supplier is also responsible for water resource management, then demand management may well be an economic alternative to the development of new sources. However, governments and municipalities are still too often tied into the traditional large engineering solutions for increasing supply. They need to take the initiative to stimulate or require demand management as an alternative to increased supply. The question then arises as to whether it is possible to legislate for or regulate and properly enforce water efficiency.

In England and Wales the Water Industry Act 1991 imposes a statutory requirement on the owner or occupier of property to install water fittings, and maintain those fittings in good repair, so that they do not lead to the waste, undue consumption, misuse or contamination of the water on or off the premises.

Many countries have building codes that set standards for construction. These often include standards for water fittings and installations. Generally these codes are legal requirements that are enforced by a public authority or government agency. In England and Wales, the Water Supply (Water Fittings) Regulations 1999 set out the legal requirements for the prevention of contamination, waste, misuse, undue consumption and the erroneous measurement of water supplied by a water undertaker. The regulations, although made by government, are enforced by the local privatised water undertakers. They include a number of requirements that are intended to promote water conservation. For example, they set limits on the flush volume for newly installed WCs and limit the volume of water allowed in a single cycle for washing machines and dishwashers. They also require that water fittings are of an appropriate quality and standard and are suitable for the circumstances in which they are used. In addition the regulations contain requirements for water system design and installation that among other things are intended to ensure that systems do not contribute to the waste of water.

Enforcement is clearly a key factor in ensuring adherence to regulations. In the UK, the Water Supply (Water Fittings) Regulations 1999 apply only at the point of installation. They cannot therefore be cited to prevent the *sale* of fittings and appliances that do not comply with the regulatory requirements. The regulations specify the appropriate standards that apply to fittings (and also to the installation work) and also require that fittings are fit for purpose. However, the main controls to prevent non-compliant fittings being installed include a statutory duty on the enforcement authority to enforce the regulations, a requirement that certain proposed installation work is notified to the enforcement authority (and work cannot commence until approval has been given or has been deemed to have been given), and a provision allowing inspection of the installation by the enforcement authority. But these controls are not comprehensive and so, additionally, the regulations make provision for the establishment of approved contractor schemes. Under these schemes approval requires a test of knowledge of the water regulations and also periodic inspection of the work. Non-compliance with the regulations carries a heavy fine. However, an approved contractor is required to indemnify the owner or occupier of property against non-compliance of the work with the regulations.

It is of course impracticable to require inspection of every installation irrespective of the size of the installation or potential risk of contamination, waste, undue consumption or misuse of water supplied by a water undertaker. Nor do the regulations prescribe how enforcement should be carried out. It is likely, however, that inspection will be limited to those installations and establishments that constitute a high risk or where the work is not carried out by an approved contractor. There should, of course, be spot inspections of a sample of other installations, ideally at the point of sale, but also at the point of installation, may also be necessary to ensure that new installations are water efficient.

8.4   CUSTOMER METERING

Most countries have some form of household metering or other charging structure for water used. However, water companies in many developing countries set low or flat rate tariffs, water rates which are subsidised by government, or provide free water. Although this is frequently in the interests of low-income customers, to maintain health and hygiene, it does tend to become the expected norm, and is frequently a politically sensitive issue, especially during local elections. However, there are severe disadvantages to a water company of a allowing zero- or low-rated tariff structure and not charging an economic rate for water:

- it does not encourage sensible use
- it does not encourage the mending of customer leaks
- the company has no incentive to install an active metering and meter replacement policy
- insufficient revenue is generated to provide a sustainable operation, maintenance and repair programme

Often, even on low tariffs, customers (both household and non-household) will vandalise or by-pass meters to save paying. Usually a review of a company's customer metering policy and tariff structure is included in the strategy development procedure recommended in Chapter 4.

Correcting the metering policy and tariff structure policy, in conjunction with other water conservation initiatives, is a major step towards reducing customer demand. To overcome adverse reaction from customers and to assuage political sensitivities, a pilot study could be designed within a water loss study programme. The study would include reading a sample of customer meters to check:

- how many meters are working and how many are stopped
- which of those not working are due to meter malfunction, deliberate vandalism, or bypassed (illegal connection)
- how accurate they are (under-registration)

Meter accuracy can be checked by installing a calibrated 'check' meter downstream of the meter on test. Companies should be encouraged to install class C or D meters. This is an international standard, referring to a highly accurate meter which uses a smaller inferential head whilst retaining the same size meter body, and which improves accuracy at very low flows. Locally made meters should be viewed with caution, as they are usually not of Class C or D standard.

Once the pilot area meter data have been analysed a sample of houses can be fitted with class C or D meters to demonstrate the difference, and to measure customer flows for the water balance calculation. Also demonstrated is the equitability of paying for water used, even if the tariff is low, and particularly in countries where water is scarce.

A tariff system on a rising scale can be introduced for non-household customers, again to encourage water conservation practices. A regular meter replacement programme should be introduced for these customers, particularly high revenue customers, to ensure that the company continues to maximise its revenue. Some companies have a policy where high revenue meters are changed every five years.

8.5   CASE STUDIES

Policy changes are best illustrated by the following case studies, which encompass all of the policy issues described in the previous sections.

8.5.1   The Vietnamese water industry 1994 [3]

In 1994 the Vietnamese water industry was entering a period of change, acknowledging water losses of 45–70% of production, and striving to reduce them.

Their new 'open door' policy has increased the pace of change by exposing companies to new markets, and to better standards and quality of materials, and by accelerating the transfer of technology. In 1994, the Ministry of Urban Construction issued an order to water companies to reduce water loss by 50% over the next ten years, and issued guidelines on how this should be achieved. The steps were:

- to review losses and identify the components
- to calculate the cost of control
- to eliminate flat rate tariffs
- to improve public awareness

The initiative was supported by the World Bank with the aim of training water company directors and senior engineers to help them develop short-term action plans and longer-term programmes which are:

- appropriate to the Vietnamese culture
- sustainable within their social and political structure

This was achieved by means of training workshops, attended by 70 delegates representing almost all of the water supply companies in Vietnam. The majority of delegates were directors, vice-directors, and heads of technical, financial, and planning departments. The primary objective of the workshops was to enable each delegate to design both short-term and long-term programmes to reduce water loss in his or her particular company. This would be achieved by bringing together delegates from different water companies, but with common problems, so that ideas could be developed from discussion groups and by example. The aim was to encourage delegates to develop their own sustainable solutions, to build on what they have rather than devise solutions which are unworkable or unaffordable. Delegates were encouraged to:

- examine the scale of water loss
- identify the causal factors
- assess the relative significance of real and apparent losses
- review appropriate tools, methodologies, and equipment to support programmes
- to reduce water loss
- design programmes which were feasible and sustainable for the Vietnamese economy, culture, and institutional organisations

Delegates were therefore encouraged to discuss openly the constraints and weaknesses (and also the strengths) of their system characteristics and their existing procedures, and to propose only those solutions or actions which could be realistically implemented. The workshop style of training course was unfamiliar to the delegates, but one which they welcomed. Previous training had consisted of formal lectures with few opportunities for discussion and no participation by the delegates.

Delegates were given water loss figures for countries worldwide, noting comparisons between developed and developing countries, the varying significance of the ratios of real losses to apparent losses, and their components. Delegates were then divided into discussion groups and asked to consider which of the components were most significant in the Vietnamese water industry. From the presentations which followed this exercise, several important points arose, which influenced the priority tasks for the action plans. It was accepted by all delegates that the main source of water loss is from illegal connections or illegal use, and from consumer meter under-registration. The points to address in an immediate action plan were therefore:

- to ensure that all consumers are metered, removing the flat rate tariff, which does not encourage wise use of water
- to stop illegal use by introducing more rigorous investigation of illegal connections, damaged and by-passed meters
- to ensure that fines are imposed – a system of public 'naming and shaming' through the media is already an option

Secondary actions would include:

- the replacement of non-working meters
- introduction of a meter purchasing policy which ensures that only meters which can accurately measure low flows are used
- checking or installing production meters to enable more accurate water loss figures to be calculated
- introduction of organisational changes to improve the accountability of the meter readers

It was agreed, however, that most institutional and organisational changes would be part of a longer-term strategy. During the workshops delegates gave the impression that now that the government had initiated a 'wind of change' they wished to be empowered to activate action plans. The workshops concentrated on the programmes to reduce apparent losses, because in most companies this is where the majority of losses occur. However, although the time for the introduction of advanced technology is still some years away, there are a number of techniques, like leakage monitoring, and some technologies, like flowmeters, insertion meters, and equipment for listening for leak noise, which were of interest to the delegates, and which are wholly appropriate to the Vietnam water industry.

8.5.2   Haiphong, Vietnam – 1997 [4]

The Vietnam Water Supply Association had acknowledged that water losses for the majority of water supply companies is unacceptably high. In percentage terms losses were estimated to be 35%–50% of water produced, and up to 70% in some urban areas. Following a national workshop on urban water supply and sanitation, conducted by WHO in 1997, The Vietnamese government requested that the principles and practices of water loss management should be applied to one of the towns in Vietnam. Haiphong, which was one of the cities participating in a WHO 'healthy cities' project, was selected as a test site.

Causes of losses in Haiphong Water Supply Company

From the author's (Farley) experience of water loss projects in Vietnam, apparent losses, such as illegal use, are exacerbated by failure to collect revenue, wasteful practices and excessive use. In Haiphong the factors which contribute to the high level of apparent losses can be attributed to both the company's management record and attitudes of the customers.

The main 'management' factor is the poor organisation of water production and distribution as a business, including:

- poor design and construction of the pipe network
- payment by flat-rate system rather than metered use, leading to a deficit of revenue against water distributed
- poor records of registered customers
- lack of regulations for water use and no enforcement measures
- loss of revenue from inefficient collection practices
- lack of cooperation and support from local authorities
- lack of budget to maintain or improve operations

The factors due to customer attitude are:

- lack of customer awareness of the value of water as a commodity
- waste and excessive use of water by customers, and from public tanks
- illegal use or theft of water
- lack of any incentive to use water wisely

A customer-use study in 1991 showed that individual customers use widely varying amounts, from 100 litres/head/day (1/h/d) to 1400 1/h/d. A water balance calculation for the area supplied by the main production plant showed an average per capita consumption of 800 1/h/d, but the company only collected revenue for 150 1/h/d. The total water loss in Haiphong in 1993 was 70% of production. There was, therefore, a high risk of water shortage and low level of service, with some customers receiving no supply or inadequate supply.

The water business management programme

It was against the background of poor management practice, lack of customer awareness, excessive customer use, and loss of revenue, that in 1994 Haiphong Water Supply Company introduced a model 'water business management programme' – an action plan to address water losses. Prior to the programme water losses were estimated at 65%, 45% of which were apparent or 'management' losses, and 20% real losses. Apparent losses were divided into two components:

- illegal use
- water used but not accounted for

  Illegal use comprised:

- illegal connections
- use of water for irrigation
- unregistered commercial use, such as cleaning vehicles and motorcycles, and noodle production

  Water used but without revenue collection was largely due to:

- water wasted from taps being left open
- excessive use compared with normal requirements
- misuse of public tanks and tanks overflowing
- non-collection of revenue – sometimes customers were unwilling to pay because of inadequate service

The physical losses were visible leakages from corroded pipes, broken joints and valves, and from house connections. The main factor influencing physical losses was the age of the pipe network.

The programme involved setting up models of good management practice in selected *phuongs* (wards). Each *phuong* in Haiphong has a discrete structure administered by its own public authority. It was considered essential for cooperation and communication that the water improvement programme followed the *phuong* boundaries. The aims of the programme were to:

- accurately measure the amount of water produced
- assess the efficiency of the business
- increase revenue to the business

  The aims of the activities to be carried out in each *phuong* were to:

- strengthen relationships between the company and the customer
- promote the responsibility of the company and public authorities to improve levels of service

- re-establish a balance between the company and the customer so that the customer receives a better level of service and the company increases its revenue
- increase the customer's responsibility by removing the flat rate payment and ensuring they pay for water used
- remove public water tanks to avoid waste

In implementing the model, the company recognised that some necessary steps would be particularly challenging. These were:

- changing the expectation of the customer from paying almost nothing for water to paying a more realistic price for water used
- raising the level of consciousness of the customer to sensible use of water
- overcoming budget and time constraints to making network improvements, educating customers, and installing customer water meters
- improving the technical knowledge, skills base, and sense of responsibility of company staff

A principal feature of the programme was the provision of a dedicated working team for each *phuong* to carry out the activities and to address the difficulties. The duties of each team were to:

- manage the pipe network from the master meter to the customer meter
- repair all visible leaks in the network and make minor repairs to consumption meters
- read customer meters to a timetable agreed with the Consumer Services Office
- monitor and update population and consumption data
- make monthly revenue collections against bills issued by the Consumer Services Office to an agreed timetable
- maintain close contact with the public authority and police to enforce duties
- address violations of water use and disconnect illegal connections
- carry out education and awareness activities
- analyse daily data and prepare reports

To support the *phuong* programme, a number of actions were necessary to strengthen the network and improve the accuracy of data collection. Meters were installed at each treatment works outlet to measure production. Installation standards were set for each *phuong*. The actions and standards were:

- installation of distribution pipelines in pavements and main lanes, with branch connections from the distribution pipes to all households
- installation of water meters at each household, at a position for easy inspection
- disconnection of redundant pipes from the trunk mains

- installation of master (block) meters at each trunk main branch connection, to monitor flows into the *phuongs*
- distribution pipes larger than 80 mm in diameter to be of cast iron with rubber joint rings to ISO 13 standard
- use of high-pressure plastic supply pipes

Review of the programme

The author (Farley) reviewed reports produced by the company during 1996 and 1997 which set out the background, aims, and action plan for the water loss reduction programme, and the progress made in the model *phuongs* since 1993.

The estimates of the water loss components and causes show that nonphysical losses were the major component of total water loss (45% apparent, 20% real). The company has also addressed the physical losses by replacing sections of the older pipework and by introducing a target repair time for repairing visible bursts and leakages. Water loss has been reduced from an estimated 65% of production prior to the programme to 42% across the company, with a measured average of 24% in the model *phuongs*. In some *phuongs* a combination of network improvement and better water use has led to some remarkable achievements – for example, 8% total water loss has been recorded in one *phuong*.

Twenty-three *phuongs* have been improved to date. A total of 65 000 customer water meters have been installed throughout the *phuongs*, serving 270 000 customers, and measuring their water use. All households now pay for water. A 30-day limit on payment of bills has led to a 98% revenue collection rate, and revenue has increased by 250%. There have been no customer complaints, and the level of service has been improved.

The example of Lam Son *phuong* serves as an indicator of performance:

- consumption has decreased from 8000 m³/day to 1500 m³/day
- all customers now receive a regular supply compared with two-thirds previously
- water pressure level has increased

The company has also introduced a system of monitoring total water consumption in each *phuong* by installing flowmeters at branch connections, to form smaller monitoring zones within each *phuong*. Analysis of zone consumption allows zones with high consumption to be inspected for excessive water use or leakage.

8.5.3   South Pacific, 1997 [5, 6, 7]

The concept of policy development was put into practice in two countries in the South Pacific – Samoa and Cook Islands – in late 1997. The author (Farley) was engaged by WHO to appraise the operating practices of the Western Samoa Water Authority, and the Waterworks of Rarotonga, Cook Islands. The purpose of each project was to:

- review operating practices and documentation relating to the water supply and distribution networks
- develop action plans for a water loss reduction programme, including advice on community education programmes and water conservation policy
- produce procedural guidelines for future water loss management operations
- conduct field training in leak detection and location techniques

The two countries have similar characteristics and local influencing factors. Both have water shortage problems due to the unusually high customer demand. In Samoa demand is 400–500 litres per head per day. The Samoans have traditionally used river sources for washing and drinking water, and the story goes that even now they have piped supplies, they still treat the supply like a river – running water continuously. This demand is almost double the design capacity of the treatment works, and treated water is augmented with raw water in part of the network to maintain supplies. Some customers are without water for part of the day, and others receive supply at very low pressure. This scenario is a flash-point for disease epidemics, both from drinking untreated water, and from polluted water entering the network through back-siphonage or seepage through leak points and broken joints. The significance of customer waste and misuse of water, as well as leakage from the pipework and illegal connections, had been acknowledged in consultants' reports of previous visits. These referred to taps left running all day, and house tanks, which are filled at night when pressure is higher, being allowed to overflow. These factors highlight the greater influence of local culture and social habits on total water loss than physical losses from the network. Because of the volcanic nature of the soil, digging trenches is difficult and some sections of pipe are laid on the surface. This makes it easier to spot leaks but brings in other problems – one enterprising shop owner in a remote village had used 50 mm galvanised water pipes as convenient supports for his shop's porch. The villagers clearly thought this was in no way connected to their loss of water supply and complained vigorously to the Water Authority – a case of unaccounted for water pipes!

A policy to revise the *customer metering policy* and introduce a *tariff structure* is under review. A community education policy is also being considered, to encourage householders to conserve water and report leaks. As with many developing countries, a shortage of funds for equipment invariably means there is no active leakage detection, and sometimes even visible bursts are not reported or repaired. Yet the importance of leak detection and repair is well recognised by the water authority engineers. One of the cornerstone actions of the strategy was therefore to:

a)   get the team some form of *transport* (sharing with other departments if necessary)
b)   develop some basic home-made *listening devices* to locate leaks
c)   provide *training* in the use of pipe location and leak detection equipment which had been supplied by consultants two years previously but never used

However, the main thrust was the empowerment of the team and the motivation brought about by being given such a high profile job.

A similar situation exists in Rarotonga, the main island of the Cook Islands. A study by Rarotonga's 'Tourism Task Force' in 1995 estimated that by 2000 water demand for the following categories of customer would be:

* visitors in resort accommodation: 1000 litres/head/day
* visitors in self-catering accommodation: 600 litres/head/day
* resident population: 330 litres/head/day
* agriculture (total) 750 000 litres/day

The reasons for such excessive use are:

* growers using drip-feed irrigation systems which are left running 24 hours a day
* domestic customers using unnecessarily high volumes of water
* a totally subsidised water supply, which does not encourage customers to reduce daily consumption or stop wasteful practices

However, tariff reviews are notoriously political, and a new tariff structure is not expected in the short term. But a programme to *inform* customers, particularly growers, of the need to stop using excessive amounts of water has been initiated. The programme included encouraging growers to develop and use *alternative sources* such as shallow wells.

A study by the New Zealand Geological Survey in 1988 had assessed the potential for large-capacity valley storage reservoirs. Although this was considered as a long-term measure, requiring large-scale investment, the Waterworks was considering increasing the upland resources to meet increasing demand. Rarotonga therefore provided an ideal pilot area for testing the process for development of a water loss strategy. The aim was to demonstrate that a sustained programme of demand management and water loss reduction could provide the extra resources to meet demand without the need for valley storage.

As in Samoa, leak detection is passive, i.e. only visible leaks are repaired. Although two leak locators (ground microphones) had been left behind by previous consultants, both had fallen into disrepair. The Waterworks operations staff had been taught no skills in leak detection techniques or equipment use. Repairs to the pipework were being held up by a lack of repair clamps. Rubber strips made from tyre inner tubes were being used to bind the leak.

In late 1997, when the WHO project took place, the Rarotonga Waterworks was suffering from a legacy of piecemeal inputs from a succession of well-meaning but uncoordinated consultants from a mixture of organisations. Because of the financial situation in the country, the Cook Islands government relies heavily on donor aid, with good but misplaced intentions, and not always implemented in consultation with the Waterworks. A paper by an independent consultant, in 1994, reviewed the work and recommendations of previous consultants. Referring to the failure of the strategy the consultant concluded that 'the situation can be reversed by a combination of local knowledge, low-cost and low-technology measuring techniques, and user conservation habits, without relying on expensive outside consultants'.

With the island experiencing its fifth drought in 15 years, the severity of water shortage was uppermost in the minds of the Waterworks and its customers alike. Flow in the streams at almost every intake was reduced to a trickle – there was no better time for introducing a practically viable and achievable water loss management strategy. From the review of previous projects, and from an appraisal of current practice, including plans to upgrade the distribution system, it was possible to suggest an action plan for the Waterworks. During the project, four Waterworks staff were assigned to a leak detection team and given practical field training in pipe location and leak detection techniques. As part of their training and skills reinforcement the team began implementing a pipe survey and asset register, as well as finding significant numbers of leaks.

Although the Waterworks was suffering from a legacy of failed or defunct plant and equipment inherited from a previous project, the basis of a sound strategy for managing the network had already begun. This included the installation of bulk meters at the intakes, a design for zoning the network and using zone meters for leakage monitoring. Although many of these meters were defunct, and zoning had never been implemented, the Waterworks engineers were familiar with the concept.

Recommendations were made for a number of improvements to current and future operating practices. Some of these are short-term and low-cost, requiring little extra investment. Other improvements are for the longer term, requiring some capital investment. Water shortages and drought measures could be addressed in the short term by an active policy of leakage reduction and demand management, and in the medium term by investment in upgrading the existing catchment intakes and providing extra reservoir storage capacity. Funds would be better invested in these projects than in deep valley storage.

Specific action plans, together with procedural guidelines, were drawn up for the Waterworks staff to follow. These were accompanied by a specification for recommended equipment to support the action plans. Most importantly, recommendations were made to WHO, as the funding agency, for a further project to support the Waterworks in their implementation of the action plans. This support would consist of further training, demonstration of techniques, and regular monitoring.

The short-term action plan (0–3 months) contained actions which could be carried out immediately, at little or no cost to the Waterworks, to start improving supplies. Examples of these actions are:

• modification of intake screens to reduce fouling by debris
• preparation of zone plans
• pressure monitoring to identify critical areas
• an active programme of pipe location and leak detection using repaired equipment and sounding sticks
• improvement of pipe repair techniques

The medium-term and long-term action plans (3 months–3 years) address elements of upgrading of the network, the repair or replacement of bulk meters, and the extension of zoning and monitoring systems. Pilot study areas to demonstrate zonal monitoring and leak detection techniques are an essential feature of this stage. Others included addressing the non-physical losses in the network and the excessive use of water by customers.

In January 1999, a little over one year after the WHO mission, WHO sent a questionnaire to the Rarotonga Waterworks to ask which of the actions had been implemented. Although some of the short-term actions had been completed (intake screens, pressure monitoring) very little of the strategy had been implemented due to lack of funding, lack of staff, lack of equipment, and lack of support. A follow-up study was arranged for the end of 1999, to re-visit the strategy and regain the motivation of the staff. However, this time there would be an added benefit of implementing pilot study areas and purchasing equipment to leave behind for the operations team to activate one of the main planks of the strategy.

In the two years since the last mission, there has been a major programme of network upgrading in the eastern sector of the island. All sub-mains and associated isolating valves and hydrants etc. in this area (but not the trunk or ring main) have been replaced. With assistance from the South Pacific Applied Geoscience Commission (SOPAC), the Waterworks is constructing a computer hydraulic model of the network. During the mission, the consultants would advise on positioning of extra pressure monitoring points, and use data loggers to gather pressure data, to assist with calibration of the model.

One of the key recommendations of the previous mission report was that the Waterworks should implement a pilot study area to demonstrate the benefits of dividing the network into discrete zones, to facilitate leakage assessment, monitoring, and detection. The report also recommended that before a pilot study area could be implemented the Waterworks would need to purchase certain items of plant and equipment. These included:

- flow meters, to measure flow rates of inputs into the zone, and to monitor leakage
- extra isolating valves, to secure the zone boundary
- data loggers, to capture flow data from the meters and pressure data from pressure monitoring points within the zone
- a pipe locator for non-metallic pipes to complement the metallic pipe locator already owned by the Waterworks
- a range of leak detection and location equipment to initiate and maintain a leak detection programme

None of these had been implemented due to lack of finance, but it was noted that this had not precluded the purchase of a suite of computer and GIS equipment and software. WHO supplied a package of equipment to the Waterworks for use by the consultants to implement the pilot study area, and to demonstrate techniques for detecting and locating leakage. The equipment is to be retained by the Waterworks at the end of the mission. The equipment comprises:

- 2 × 100 mm Kent Helix 4000 flowmeters for zonal flow monitoring
- 1 × 150 mm Kent Helix 4000 flowmeter for zonal flow monitoring
- 3 × dual-channel Wessex data loggers for flow and pressure data capture
- 1 × Radiodetection RD500 pipe locator for non-metallic pipes
- 1 × Palmer Mk5 ground microphone system for leak location

In addition WHO has purchased a Palmer MicroCorr 6 leak noise correlator, which is to be kept in an equipment 'pool' in Manila, for loan to consultants on similar missions. The correlator was loaned to the consultants for the duration of this mission, so that the Waterworks staff could be trained in its use, and could assess the suitability of the instrument for purchasing for future leak detection activities.

Rarotonga Waterworks plans to upgrade its supply and distribution network over the next 5–6 years. 24 km of sub-mains have been replaced in the eastern sector of the network. By closing isolating valves on the ring main and installing a flowmeter to measure exports into an adjacent area, a discrete zone will be created. The zone is supplied by trunk mains from three source intakes, Tupapa, Matavera, and Turangi. The work was carried out by Waterworks engineering and operations staff. Zoning the network was particularly recommended by WHO consultants during an advisory visit in October 1997.

Supported by the Ministry of Works, Energy and Physical Planning (MOWEPP), the Waterworks has drawn up a proposal to Asian Development Bank (ADB) for funding for a programme to reinforce the remainder of the network. The programme is in several stages, phased over 5–6 years:

- Network upgrading (years 1–3). The remainder of the sub-mains (84 km) will be replaced and flowmeters installed at each source intake. Zoning the network will allow improved leakage monitoring and assessment to take place, and will guide ongoing leak detection activities. The programme will also include a pilot study to assess customer demand to guide future charging policy.
- Intake upgrading (years 4–6). Improved screening, intake modification and upgrading, reservoir storage.
- Treatment (year 5). Design and installation of filtration and disinfection plants
- Groundwater development (year 6). Development of boreholes at selected sites.

8.6   REFERENCES

1   United Nations Human Development Report (1998) *World Urbanisation Prospects 1996 Revision.* New York: UN.
2   Edwards, K and Martin, L (1995) 'A methodology for surveying domestic water consumption', *CIWEM Journal*, 9, October, 477–88.
3   Department of the Environment, Welsh Office (1992) *Using Water Wisely: Consultation Paper.* London: DoE.
4   Farley, M (1995) 'Reducing water losses in Vietnam'. *Proceedings, of 21st WEDC Conference 'Sustainability of Water and Sanitation Systems'.* Uganda: WEDC.
5   Farley, M (1997) Mission Report RS/97/0705 Socialist Republic of Vietnam. Manila: Regional Office for the Western Pacific of the World Health Organization.
6   Farley, M and Dakin, M (1997) Mission Report RS/97/0627 – Cook Islands. Manila: Regional Office for the Western Pacific of the World Health Organization.
7   Farley, M and Bull, G (1999) Mission Report MR/1999/0784 – Cook Islands. Manila: Regional Office for the Western Pacific of the World Health Organization.
8   Farley, M (1997) Mission Report RS/97/0102 – Western Samoa. Manila: Regional Office for the Western Pacific of the World Health Organization.
9   Farley, M and Dakin, M (1996) Mission Report RS/96/0112 – Western Samoa. Manila: Regional Office for the Western Pacific of the World Health Organization.

9

Ensuring sustainability

Previous chapters have dealt with developing a strategy for a water loss management programme. Some of the elements of this strategy require bringing the network up to a sufficient level for the techniques for assessment, monitoring and control to be put in place. Some of those techniques are programmes to address the identification of apparent losses – these may require social, cultural, political, and legal changes, and a longer term view. Other techniques are needed to address real losses – leakage detection and location programmes, purchase of equipment and network strengthening. These can be implemented in the short term, provided the required financial support is forthcoming.

What is common to all programmes is the need to maintain the motivation of staff, the understanding of what is being achieved, and the provision of skills and technology to sustain the programmes. This chapter deals with the issues which ensure that advances made in introducing a water loss strategy are sustained. These are:

- ensuring appropriate staffing levels
- staff education and training
- operation and maintenance (O&M)
- assessing and monitoring performance

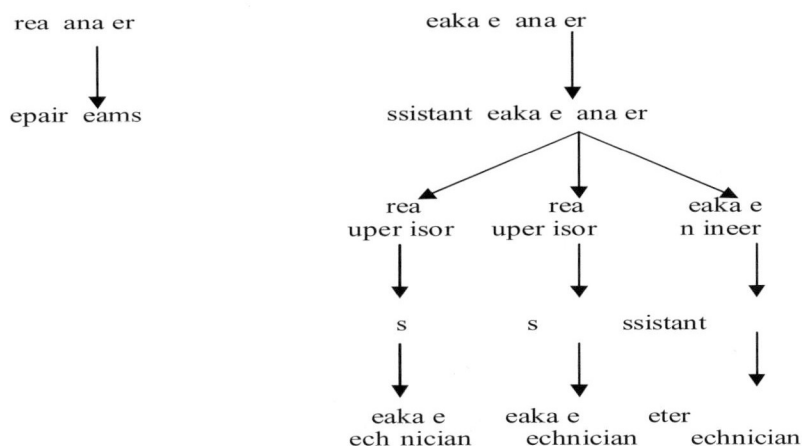

Figure 9.1 A typical leak detection team.

9.1   STAFFING LEVELS

A leakage management programme will succeed only if sufficient weight is attached to staffing a leakage team. The team should be:

- sufficient in numbers to carry out all the tasks required in the company's geographical area
- skilled enough to carry out the tasks
- credited with sufficient responsibility
- motivated enough to perform to set targets and criteria

A typical team structure, based on a two-man leak detection unit (LDU) per 30 000 connections, and a company area of 300 000 connections, is shown in Figure 9.1.

The structure is based on the construction favoured by many companies, where the operations 'arm' is separated from the maintenance and repair facility. Team responsibilities are as follows:

- Leakage manager – an experienced engineer, who has overall responsibility for the leakage team and its performance
- Assistant leakage manager – an engineer trained in leak detection technology, to act as a go-between for field and office-based staff
- Area supervisors, each responsible for 5 LDUs
- LDUs, each containing 2 leakage technicians, responsible for leak detection and location
- Leakage engineer, responsible for monitoring, support service and guidance for the leak detection teams

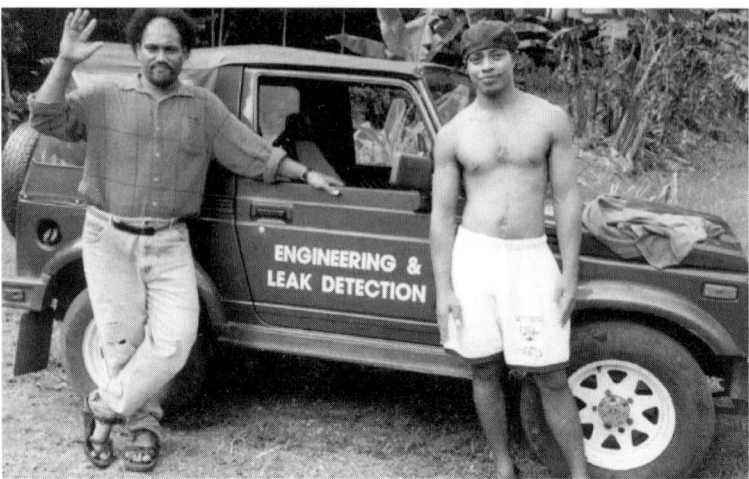

Figure 9.2 Motivating and empowering staff.

- Assistant leakage engineer, providing a supporting role to the leakage technician in the field
- Meter technician, responsible for meter checking and repairs

The LDU would be also be responsible for surveying the pipe network, both to locate and re-map the pipework in addition to detecting and locating leaks. A reasonable output for a two-person team would be between 2 and 4 km per day, depending on conditions (e.g. urban or rural) and it may be advisable to strengthen the LDU with two extra leakage technicians to do this work. Once a pipe survey and a 'first pass' leak survey has been completed, it should be possible to keep leakage under control with two technicians, particularly if the network can be divided into leakage monitoring areas (DMAs).

Provision of transport is essential, and the very minimum requirement is a dedicated van, solely for the use of each leak detection unit. The transport should be a covered van, fitted out with racking so that equipment can be carried safely and out of sight.

The van should be marked with the company logo, and with a legend, for example – 'Leak Detection Team' (see Figure 9.2) painted on the sides. This serves three purposes:

- to give the technicians a sense of pride and purpose
- as a safety measure, to identify the water company and advise residents of its purpose, particularly during night work

Losses in Water Distribution Networks
A Practitioner's Guide to Assessment, Monitoring and Control

Erratum
Due to a printing error, Figure 9.1 (page 207) has been reproduced incorrectly.
IWA Publishing apologises for this error and the corrected figure appears below:

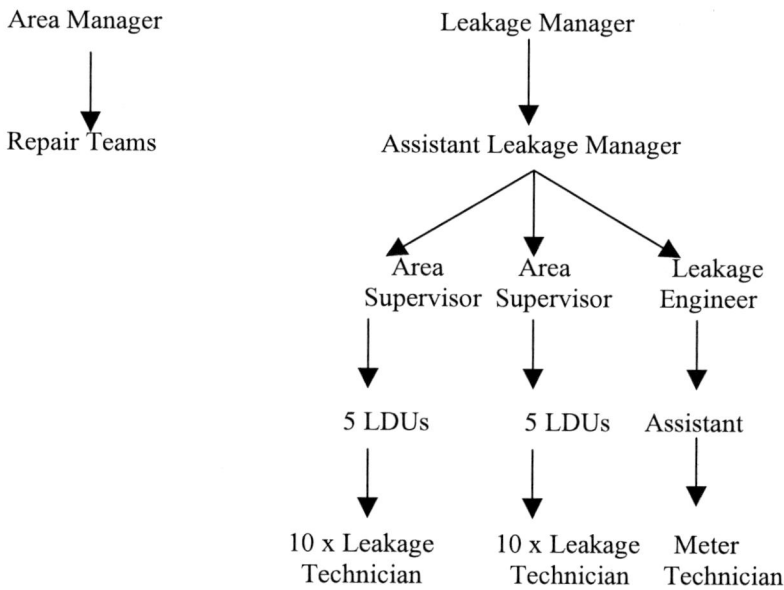

Figure 9.1 A typical leak detection team

- to raise the company's profile, and to make consumers aware of how the company is addressing leakage

9.2   EDUCATION AND TRAINING

One of the essential tools of a strategy, once in place, and a key issue for sustainability, is the training of local staff in the skills and techniques of water loss management. Training encompasses the motivation of staff, transfer of skills in the techniques and technology of leakage management, and system operation and maintenance. The following sections describe the philosophy and approach for meeting those expectations, and for ensuring a sustainable leakage management strategy.

Skills transfer and awareness training is required at an early stage, to appraise all staff of new concepts and methodologies, followed by continual checks and monitoring. There should be a 'top-down' approach, where key decision makers and senior managers are first appraised of the principles of the water loss programme, and their commitment and motivation is cascaded down to other staff. The following sections illustrate the skills and technologies required.

9.2.1   Training and skill transfer requirements

Elements include:

- measurement and estimation of water loss and leakage
- development of a water loss management strategy
- design and installation of leakage monitoring districts
- analysis and interpretation of leakage data
- detection, location and repair of leaks
- planning rehabilitation of pipework
- operation and maintenance of plant and equipment

9.2.2   Approach

There is a need to address the tasks, the problems and the constraints associated with introducing a leakage management programme at all levels of the organisation. It is important that an understanding of the principles of the programme, the steps in its design and implementation, and support for the tasks involved, filters down from senior management level to operations level. A training programme will therefore include *awareness seminars* for senior staff (and also to raise public awareness) *training workshops* for engineering and technical staff, and *continuous practical training* for operations staff.

Awareness seminars

A carefully planned and targeted awareness seminar is the first step in the training programme. The main aim of the seminar should be to brief the senior managers and decision makers on the aims of the project programme, the steps required to implement it, and the plans for staff training. Once perception is raised at the top, motivation will be enhanced at all levels of the organisation. These delegates will be given an overview of the issues and methodologies to appraise them of the objectives and cost benefits of a leakage management policy and the steps required to introduce one. They will clearly be more interested in the financial and institutional aspects of the programme and in the benefits of increased operational efficiency, but will also require an overview of the technical aspects.

Training workshops

The aim of these is to ensure that within the constraints of the system, recommendations for best practice are made, and skills are transferred to engineers and technical staff responsible for managing and controlling leakage. It is unlikely that one workshop will be suitable for raising the awareness of the whole range of an organisation's staff. The workshops should therefore be tailor-made for each category of staff: for example, engineers, while still requiring an awareness of the management issues, would benefit from those modules with a practical and technical approach. Technical and operational staff would be given an awareness of the programme stages and requirements, and the hands-on skills training which will follow. Workshops for operational staff would be modular, i.e. they would cover particular aspects of leakage management, such as metering, flow measurement, pressure control, data logging etc. as a precursor to the field training programme.

Continuous practical training

This element is for operational staff responsible for leakage management in the field. The training would be structured in modules, each module covering a particular task or skill, to reinforce workshop training. Training in the field requires a long term view, with continuous repetitive exercises.

Additionally, it is important to integrate the continuous training element with the ongoing leakage management programme, so that trainees are working alongside company engineers, contractors or consultants for certain periods at selected times. This on-job training will result in efficient skill transfer and an understanding of day to day operational tasks with the accompanying problems and solutions.

Figure 9.3 Vietnam national workshop, Ho Chi Minh City, December 1994.

Seminar and workshop programmes and the continuous practical training programme should be designed with the needs of each group of staff and their respective requirements in mind.

Appendix E contains example contents for each programme.

9.2.3   A national workshop

A precursor of water loss studies in countries where the procedure is unfamiliar is often a national or regional workshop for water engineers and directors who can themselves become 'trainers' The case study on the Vietnamese water industry [1] in section 8.5.1 and the illustration of a national workshop in Figure 9.3 provides an example of this. The workshop is based on World Health Organization guidelines for leakage management [2] and would focus on the design, preparation and implementation out of water loss management and leakage control projects. The workshop content is similar to that for the company-based workshop, described in Appendix E, but would be enriched by the experiences of engineers from different areas and companies. The initial target audience for the workshop in each region or country is expected to be a core group of engineers, who will derive their own courses, workshops, and seminars, and who will disseminate the methodologies to other practitioners in the water supply sector.

Water loss and leakage control activities are grouped into workshop modules. The criteria for grouping activities take into account the types of professionals involved in their execution (managers, engineers, technicians, etc.) and the characteristics and level of development of the water distribution network. Each module is self-contained so that the modules can be directed to different types of target populations. Therefore, the modules contain as much information as required for the achievement of their particular objectives, and to provide a high degree of flexibility in their assembly in preparation for training courses.

The workshops adopt a logical and 'user-friendly' approach to training water practitioners at a range of levels, from senior managers to leak inspectors. Each module can be varied in content depending on the depth of knowledge required for a particular level of trainee. For example, engineers and managers could explore in detail the institutional and financial aspects of leakage control, and would benefit from a cost–benefit exercise to select and develop an appropriate policy. Engineers and technicians responsible for managing a system and detecting leaks would benefit from an understanding of these principles, but the main thrust of their programme would be based on those modules with a more practical and technical approach to system management.

Two essential components of this type of workshop are:

• *tailoring* the course to a particular utility or community's requirements by finding out current system practice, problem areas, strengths and weaknesses etc., and then involving the trainees in producing an 'action plan' for their system, based on local knowledge and new skills gained from the workshop
• having a *practical demonstration* of the available equipment over a range of technology, followed by field demonstration (at a pre-selected and prepared site near to the workshop venue). All trainees should have the opportunity to handle the equipment and become familiar with procedures (e.g. programming a data logger, measuring a flow profile, locating a leak)

This approach requires a little planning beforehand and during the first day of the course. Delegates are required to provide written information about their water supply, e.g. physical details like topography, population and demography, cause and magnitude of losses, etc. as well as a description of the current leakage control policy, if any. At a suitable point during the first day, when the workshop programme, etc. has been introduced, individual trainees, or a representative of a group, are invited to make a brief presentation on the background and current practice of their water supply department. This has three purposes:

a)   it acts as an icebreaker
b)   it helps to stimulate discussion
c)   it provides local material and experiences

From this point on, all the points raised can be dealt with in subsequent modules, again with trainee feedback and active participation. The local knowledge thus gained is also invaluable for constructing an action plan at the end of the workshop, when the trainees will benefit from group work – comparison of different ideas and views will act as an additional stimulus to discussion. This philosophy is also explained to trainees during the introduction.

It is important to emphasise during the introduction that a water loss control programme can be initiated in any water undertaking, even in those with intermittent supply (supply can be partially restored by a low-activity policy such as repairing visible leaks in overground pipes – a policy practised in many developing countries as the first stage of their leakage control programme). It is the sustained activity which is important, i.e. building an appropriate, achievable and sustainable water loss management strategy.

9.2.4 Publicity and profile

It is good for the water company, its customers and shareholders (as well as its critics) to publicise the results and benefits of new projects. It is suggested that at some point during a water loss project a national, regional or local seminar is held to raise public and political awareness of the organisation's goals and achievements. Timing will depend on project milestones, but a suitable time would be at the end of the project, when the strategy has been implemented, data have been gathered, and results achieved.

Customer services staff also have a part to play in raising the profile of the company. Training requirements for these staff focus on the interaction between the company and the customer, so that customers are involved in the successes of the project. A training programme could include making staff not directly involved in leakage management (e.g. call handlers, billing staff) aware of:

- the significance of leakage
- its impact on the customer
- the steps being taken by the company to reduce it
- targets, successes and trends (results so far)
- how to brief customers

The role of the call handler is particularly important in promoting the company and in fostering good customer relations. Call handlers should be aware of the steps to follow:

- what to tell the customer if asked about leakage
- the route to ensure the problem is passed to the leak detection team
- how the customer can help by describing the type of leak and its situation

The training method should be by means of a short (2 hours) informal workshop, which is informative and interactive. A typical programme would contain:

- driving forces and external pressures on the company – meeting the demands of customers at the same time as running a business
- how leakage has become a major issue
- a brief look at leakage – cause, effects, how it is measured, how much etc.
- the company's viewpoint – long-term investment, short-term actions
- a brief look at the methods and equipment for leak detection
- strategies and results so far, comparisons with other companies
- what the customer sees – the anomaly of water shortages and high leakage

9.3   OPERATION AND MAINTENANCE

Operation and maintenance (O&M) is crucial to the successful management and sustainability of water supply networks, whatever the level of technology, infrastructure, and institutional development. The O&M philosophy applies as much to hand-pumps as it does to more advanced treatment works and water distribution systems. It requires forward planning and technology transfer at all stages from installation of plant and equipment, through operator training and hand over, to routine operation and upkeep. O&M therefore encompasses equipment selection, spares purchasing and repair procedures as well as best practice in operating and maintaining the system. Figures 9.4 and 9.5 illustrate poor O&M practice.

Inadequate O&M leads to inefficient practice, ineffective services, and waste of precious resources. The key factors which have been found to contribute to inadequate O&M procedures [3, 4] include:

- low profile of O&M
- inadequate access to information

Figure 9.4 A leaking hydrant.

Figure. 9.5. A leaking transmission main joint.

- insufficient funds and misuse of funds
- inappropriate design
- poor management and overlapping responsibilities
- political interference
- lack of policies and legal frameworks

These factors are exemplified in the case study described in 9.3.1.

9.3.1   O&M case study

During a study in Zanibar in 2001 [5] the O&M practices were assessed – the following factors, typical of developing countries, contributed to poor O&M:

- lack of a water policy
- no customer charging policy, therefore no regeneration of funds for O&M
- low profile of O&M and lack of perception of the need for O&M by decision makers
- lack of budget for maintenance, materials, spare parts, tools and equipment
- poor design of plant and equipment
- lack of consultation with local operators, and resulting inappropriate choice of technology, by donor organisations
- lack of spares, training and follow-up for operation and maintenance of plant installed by donor organisations
- lack of vehicles for on-site O&M
- lack of fuel for vehicles and plant
- failure to repair water pipe breakages, increasing the risk of cross contamination from sewage and septic tank effluent water entering the distribution network

untreated and without disinfection. While the majority of sources are springs, artesian wells, boreholes and tube wells, with a relatively low risk of contamination, there is a danger of cross contamination in the distribution network

Zanzibar lacks a coherent water policy for cost recovery and revenue generation. While water resources are plentiful, customers are rationed (supplied one day in three) in parts of the northern island of Pemba due to a poor distribution network and low pipework maintenance. There is also no water demand management or water conservation policy in a community which has increasing demand from a rising population, haphazard housing construction and waste of water. However some successful community projects have demonstrated the benefits of community/ government partnerships, promoting pride of ownership and motivation for good O&M practice.

Despite the opportunities for rainwater harvesting (abundant rainfall, corrugated-iron roofs on many buildings), there was little evidence of this alternative source. One school visited had developed a system of rainwater collection via roof-guttering into a low-cost ferro-cement storage tank, but the site chosen was unfortunately in a low-rainfall area and the facilities are unused.

Installations by donor organisations, although gratefully received, sometimes provide difficulties for O&M. Examples are:

- lack of consultation on design of plant, and on choice of manufacturer of equipment and instrumentation the resulting choice of materials and technology is inappropriate to local skills, repair or replacement parts when equipment breaks down
- lack of training in O&M of newly installed equipment
- lack of a process for contractor accountability and follow-up when plant fails to work. At one hospital visited a solar panel for the hot water system had never worked, but neither had the installers been called to account

9.4   ASSESSING AND MONITORING PERFORMANCE

Once targets have been met, the key issue is then the ongoing maintenance and management of leakage at the reduced level. All aspects of leakage management requires constant effort if leakage is to be kept in a reduced state. Leakage never goes away, it is something which requires constant attention, otherwise leakage levels will rise and could return to the rate before the leakage reduction programme, wiping out much of the work and investment. If anything, leakage management becomes more difficult as this stage than during the reduction stage. The focus of attention on a major investment project has passed, further investment may be more difficult to obtain, and the ongoing work is seen as a cost burden which does not produce any benefit to the organisation.

An efficient and effective set of procedures must be put in place during the leakage reduction plan to ensure that once a set target has been reached, the leakage level is maintained at or near that target in future years. Leakage is like a spring, and unless downward force is maintained, it will bounce back. These procedures apply at three different levels:

- strategic
- tactical
- operational

Each of these is considered in turn.

9.4.1 Strategic monitoring

The overall indicator of performance on leakage management comes from the annual water balance calculation. However, the water supplier should not wait 12 months between carrying out these calculations. It is important that trends within the year are monitored, and that corrective action is taken if it appears that the annual target will not be met. The situation is similar to the financial management of a company, which has to file annual company accounts. Profit and loss figures will be produced more frequently to ensure the company is meeting monetary targets. It is recommended that water balance calculations be carried out every quarter, and possibly monthly where targets are critical or in areas where changes are being made to the regime which has applied in previous years.

Water balance calculations rely on accurate monitoring of distribution input (equivalent to income), and all elements of water use (equivalent to expenditure). The estimate of water loss is based on the difference between distribution input and water use, and is the equivalent of profit being the difference between income and expenditure.

9.4.2 Facilities monitoring and maintenance

The facilities which have been established during the reduction stage have to be maintained. This work tends to involve periodic inspections:

- district meters have to be calibrated or checked at regular intervals
- district boundaries have to be maintained
- statistics such as property counts have to be kept up to date
- pressure reducing valves will require monitoring and maintenance
- equipment such as correlators will need regular calibration and periodic replacement

Records should be kept for each facility, and piece of equipment, similar to vehicle maintenance records, to show what maintenance work has been carried out and when. It is useful to maintain a file package for each DMA containing all relevant information which is updated with the results of the most recent survey work, changes to boundary valve positions, etc. Computer packages are available to store all this data electronically, possibly linked to the digitised mains records.

9.4.3   Operational monitoring

Day-to-day management of leakage is a painstaking task, which requires ongoing monitoring of large volumes of data and information. Computer systems are available to store meter readings, pressure data, consumption data etc., which produce leakage values for DMAs based on nightline information or regular (e.g. weekly) readings. These systems can also be developed to prioritise areas for leak detection surveys.

While such systems help with the task of organising leakage management operations, they also require maintenance.

A key element of the ongoing monitoring is the assessment of productivity from the leak detection staff. During the reduction stage, it is possible to monitor expenditure in terms of the cost per Ml/d saved. However, when maintaining leakage at a set level then this is more difficult. Other measures have to be used, some of which are similar to those used in the reduction stage.

Performance monitoring should include consideration of the following key issues:

* volume of leakage found
* safety
* productivity
* quality and reliability
* customer contact

    Typical performance measures will include:

* hours per leak detected – sometimes converted into equivalent service pipe bursts (espbs)
* properties surveyed/operator hour
* number of private leaks investigated per day
* number of abortive excavations (dry holes)
* number of abortive visits by water company staff
* adverse customer reaction
* adverse client reaction

Some companies use a points system with points being allocated each week or each month according to a number of factors including leaks found (more points for mains bursts than service bursts and fittings leaks), other defects reported (e.g. defective valves), time-keeping, presentation etc. A league table may be used, particularly when employing contract staff, showing the relative performance of each operator or each team employed. This comparative performance acts as a natural competitive incentive. Additional motivation can be applied by awarding more work to contractors who top the league table, and less to those who come at the bottom. Minimum standards may be set and regular poor performance may result in a contract being terminated or not being renewed.

Staff should be appraised by regular interview, and the other management techniques described in section 6.7 should continue to be applied to ensure that standards do not drop once the goal of reaching the set leakage target has been achieved.

9.4.4 Use of new technology and operating practices

During the leakage reduction phase, the primary management incentive is achieving the leakage target. Once it has been achieved and maintained for a period (say 1 to 2 years) the next strategic aim should be to continue to achieve the same level of leakage at an ever-reducing operational annual cost. This will require investment in research and development and use of new technology, in order to gradually reduce staffing levels. Efficiency savings come from:

- Use of techniques which make ongoing ALC operations more efficient, such that the same result can be achieved for less effort.
- Review of data and assumptions which go into the leakage calculation, whether these be based on night flow estimates or annual water balances. In many cases, some of the demand which is initially thought to be leakage turns out to be hidden customer use, operational use, or is due to meter inaccuracies.
- Review of practices and staff levels to cover seasonal variations. The tendency for leakage to increase is often a seasonal phenomenon, but it is common for the same number of staff to be employed all year round. By investigating seasonal variations, it is possible to make savings by making changes to working practices, and duties at different times of year, or by supplementing a permanent resource with contracted staff for a few weeks at a time when needed.
- Some companies monitor weather forecasts and use the data to predict the number of bursts which are likely to occur in the next few days. From this they decide how many leak location and repair staff to have on standby and for call out arrangements.

9.5   REFERENCES

1   Farley, M (1997) Mission Report RS/97/0705 Socialist Republic of Vietnam. Manila: Regional Office for the Western Pacific of the World Health Organisation.
2   Farley, M (2001) *Leakage Management and Control: A Best Practice Training Manual*. Geneva: World Health Organization.
3   Farley, M (1995) 'Operation & maintenance – the WHO toolkit'. *Proceedings 21st WEDC Conference 'Sustainability of Water and Sanitation Systems*. Loughborough: WEDC
4   World Health Organization (1994) *Operation and Maintenance of Urban Water and Sanitation Systems: A Guide for Managers*. Geneva: World Health Organization.
5   Farley, M (2003) 'Operation & maintenance: the starting point for improving efficiency of water supply and sanitation services in developing countries – Zanzibar case study'. Paper presented at 2nd International Conference on Efficient Use and Management of Urban Water Supply, Tenerife, 2–4 April.

Case Study 1

An evaluation of the water distribution system for system losses in Sarina Shire Council, Australia

BACKGROUND

An analysis of the water consumption data and system pressure was undertaken to assess the efficiency of the Sarina Shire Council's water distribution system. The study was carried out by Wide Bay Water Corporation of Hervey Bay, Queensland. Their report details the current water losses and indicates the potential water savings, financial benefits and improvements in customer service which can be delivered by implementation of a programme of leakage control, district metering and pressure management.

The assessment process for the project was undertaken using the concepts and techniques recommended by the International Water Association (IWA) Task Force. The Water Services Association of Australia (WSAA) Benchloss Software was used to calculate real losses and associated performance indicators.

© 2003 IWA Publishing. *Losses in Water Distribution Networks* by Malcolm Farley and Stuart Trow. ISBN: 1 900222 11 6

The project scope includes:

- quantification of the amount of water loss attributable to leakage and high pressure
- a water balance
- benchmark the water losses against the infrastructure leakage index (ILI)
- recommendations for the sectorisation of the water supply network
- highlight the economic benefits of addressing system losses
  - defer the need for costly water upgrades
  - reduce the amount of unbilled 'lost' water
- recommendations for a wide range of strategies that will increase overall system performance

Benefits of water loss management

The benefits from the proposed water loss reduction and management programs were given as follows:

- Short-term financial benefits are associated with the costs of water treatment and delivery and include savings in chemical treatment costs, sludge disposal costs and power costs.
- Longer-term benefits relate to whole-of-life asset costs and include a reduction in pipe failures, extended asset life and savings in the costs of repairing burst mains.
- Indirect financial benefits also occur with more efficient use of existing water supplies. In particular, reduced water losses help ensure that existing water supplies can meet future increases in demand. This can defer construction of new water infrastructure such as treatment plants and mains.
- A degree of drought security. In the event of droughts the security of supply can be maintained for longer periods.
- Increased knowledge of the distribution system. This enables staff to become more familiar with the system, including the location of mains and valves. This knowledge assists utilities to respond more quickly to emergencies such as mains breaks and provides an early indication of any increases in water losses from leakage.

Improved public relations can be expected as the council can inform customers of their efforts to save water, save money and improve service delivery. Field teams undertaking water audits, leak detection and maintenance work will also provide visual evidence that the water system is being well maintained.

Table CS1.1 Water supply data for Sarina Shire Council.

	Sarina Township	Northern Beaches	Armstrong Beach
Water supply input to the system	313.2Ml	288.2 Ml	51.1 Ml
Number of water supply connections	1282	1490	249
Number of properties (commercial and domestic)	1282	1490	249
Authorised consumption	270.5 Ml	280.4 Ml	47.0 Ml
Length of mains	48.8 km	74.3 km	7.3 km
Production cost of water	AUS$150 000	AUS$90 000	AUS$27 000
Average selling price of water	AUS$1100/Ml	AUS$1100/Ml	AUS$1100/Ml
System pressurised % of time	100 %	100%	100%
Unit value of water supplied	AUS$600/Ml	AUS$600/Ml	AUS$600/Ml

Findings – current water use

From the initial data supplied by Sarina Shire Council for the period 1 July 2001 to 30 December 2001, the findings showed that demand was 652.5 Ml.

The data for each supply area is shown in Table CS1.1.

From the data supplied it was determined that there are financial benefits for each of the supply areas in Sarina Shire Council, achievable by addressing water losses as outlined in the report. These are shown in Table CS1.2.

If the average selling price of water is AUS$1100/Ml then the 54.6 Ml (total of all supply areas) of non-revenue water represents a potential revenue loss to Sarina Shire Council of AUS$60 060 per half-year period.

Per connection consumption was calculated as:

* Sarina Township   – 1326 litres/connection/day
* Northern Beaches – 1051 litres /connection/day
* Armstrong Beach – 824 litres /connection/day

Table CS1.2 Water loss data for Sarina Shire Council.

	Sarina Township	Northern Beaches	Armstrong Beach
Value of real losses	AUS$25 300	AUS$4600	AUS$2400
Real losses (estimated)	42.1 Ml/half year 178.5 l/conn/day	7.7 Ml/half year 28.1 l/conn/day	4.1 Ml/half year 89.5 l/conn/day
% of the system input.	13.4%	2.7%	8.0%
Non-revenue water (estimate)	42.7 Ml/half year	7.8 Ml/half year	4.1 Ml/half year
NRW production cost (half year)	AUS$25 600	AUS$4700	AUS$2400

System water losses were:

- Sarina Township: 42 Ml or 178.5 litres /connection/day (13.5% of system input)
- Northern Beaches: 7.7 Ml or 28.1 litres /connection/day (2.7% of system input)
- Armstrong Beach : 4.1 Ml or 89.5 litres /connection/day (8.0% of system input)

Comparison of water losses

Real losses cost equates to 8.3% of the total volume of water being put into the system for the period July 2001 to 31 December 2001. Figure CS1.1 shows the current standing of each of the Sarina Shire Council's supply areas with respect to benchmarking comparison to other international water authorities.

Apparent losses

It was difficult to accurately gauge the apparent losses for the Sarina Shire Council distribution network, however some data was available. This included hose down usage of 0.5 Ml and mains flushing of 0.6Ml. These figures have been applied to the water loss calculations.

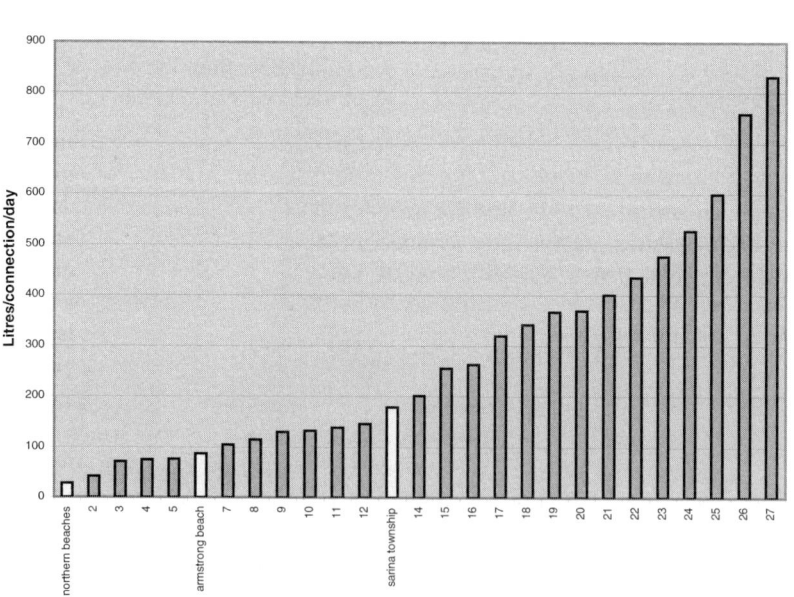

Figure CS1.1 International comparison of real losses for Sarina Shire.

Table CS1.3 ILI calculation.

	Sarina Township	Northern Beaches	Armstrong Beach
Average density of connections per km of mains	26.3	20.1	34.2
Average operating pressure (estimated)	60 m	52 m	50 m
The infrastructure leakage index (ILI)	2.00	0.32	1.32

Infrastructure leakage index: the Sarina Shire Council's ranking

The recommended International Water Association (IWA) detailed performance indicator for real losses is the infrastructure leakage index (ILI). This is the ratio of current annual real losses (CARL) divided by unavoidable annual real losses (UARL).

The IWA formula for UARL takes into account the system-specific factors of density of service connections (per km of mains), average operating pressures and location point of metering (or consumption) relative to the edge of the street. It is a non-dimensional performance indicator of the current overall management of the infrastructure for leakage purposes. The infrastructure leakage index for each of the supply areas is shown in Table CS1.3.

Figure CS1.2 shows that the three areas had relatively low leakage being ranked 1st, 5th and 12th in the data available to IWA. The more the ILI exceeds 1.0, the greater is the opportunity for further management of real losses. This is accomplished through infrastructure management and maintenance, intensive active leakage control and improved speed and quality of repairs.

The effect of pressure management should be assessed separately from the ILI calculation. A simple initial assumption for calculation is that real losses in large systems will increase and decrease linearly with average pressures, over a small range of pressure levels.

Financial estimates

In financial terms, the total volume of non-revenue water for Sarina Shire Council is estimated to be 54.6 Ml for the period 1 July 2001 to 31 December 2001. This represents a production cost to the council of AUS$32 760 at AUS$600/Ml.

Alternatively if the average selling price of water is AUS$1100/Ml then the 54.6 Ml of non-revenue water represents a potential revenue loss to Sarina Shire Council of AUS$60 000.

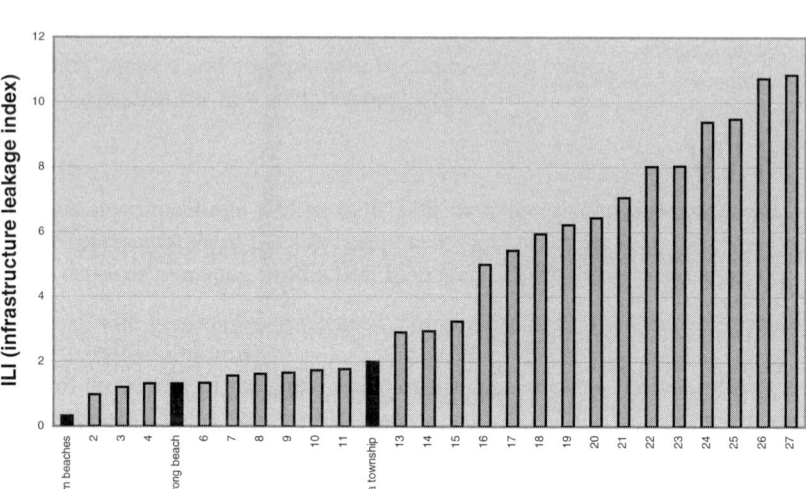

Figure CS1.2 ILI Comparison.

Potential savings and payback

The immediate real loss savings for the Sarina Shire Council network with respect to potential real loss savings should be directed at Sarina Township and to a lesser extent, Armstrong Beach where the potential for making any savings are greatest. If the losses can be reduced to a level similar to that of the Northern Beaches area of 28 litres/connection/day the potential real loss savings would be as follows:

- Sarina Township 178 – 28   = 150 litres/connection/day
                                = financial saving of approx AUS$42 000/year
- Armstrong Beach 90 – 28   = 62 litres/connection/day
                                = financial saving of approx AUS$3380/year

Hence, a potential saving of AUS$45 000 is theoretically possible. However, in actual practice, only a proportion of the potential savings are achievable. Therefore, if it is considered that 60% of potential water savings can be achieved in practice, this would equate to an annual financial saving of AUS$27 000 based on a delivered water supply cost of AUS$600/Ml. The payback period from implementing improvements to the water supply system as recommended is calculated to be 2.0 to 2.5 years based on the following cost estimates.

- Cost of district meters   AUS$15 000
- Cost of installation   AUS$32 000
- Cost of non-return valve   AUS$400
- Cost of pressure-reducing valve (fixed outlet)   AUS$900
- Leak detection survey*   AUS$16 000

*The leakage detection expenditure above does not include repair work and the cost of a leak detection survey is based on the entire Sarina Shire water network.

RECOMMENDATIONS

After an examination of the data supplied by Sarina Shire Council and on-site data collection from the supply system, Sarina Shire Council were recommended to adopt the following actions:

Phase 2: Sub-division of the Sarina Shire Council network. This involved the establishment of 10 district meter zones and installation of 14 new flow meters, 2 replacement meters in the distribution system and pressure control at Timberlands.

Phase 3: Quantify leakage and water losses in each zone by undertaking further pressure and night flow monitoring. The installation of district meters, as recommended above, to improve system performance and provide for a more accurate source of data for night flow and water loss assessment.

It was recommended that initial investigations be carried out in the Sarina Township area and Armstrong Beach area. Data analysis indicates that these two areas have the greatest amount of water losses and therefore the highest potential for reducing losses. It was also recommended that pressure control be implemented in the Timberlands district meter zone.

Phase 4: Undertake an active leakage control programme.

Water losses, pressure management, levels of service

Based on the pressure logging carried out it was identified that the network at Timberlands could be pressure reduced. The pressure reduction would not so much be for reduction in leakage, but to improve the level of customer service. Timberland is the first supply area from the pumps at Alligator Creek pump station. As the reticulation is fed directly from the pumping mains, pressures of 60 m are experienced when pumps are running. A drop of some 10 m is experienced in the area once the pumps go off at which time the area is gravity supplied from the Mt Griffith Reservoir.

As a result of this large difference in pressure, customer satisfaction was compromised and complaints the usual outcome. With the installation of a pressure-reducing valve it would be possible to maintain the pressure within the Timberlands reticulation at a constant level.

From the logged pressure data it could be seen that there is a potential pressure reduction of some 15 m to 20 m available without compromising the standard of service.

Establishment of district meter areas – sectorisation

As with most Australian water networks there is insufficient metering within the Sarina Shire network. System metering provides the following benefits:

- knowledge of where water is coming or going
- enables prompt identification of problems within the network such as leakage
- provides information on the approximate location of the leak
- increases awareness
- reduces running time of the leak

To account for distribution losses in the Sarina Shire Council water distribution network and monitor flows, the network has been sub-divided into separate smaller zones known as district meter areas (DMAs). This arrangement is more manageable for investigating and controlling of water losses within the network. It will also identify where the greatest losses or usage are and thereby highlight which areas should be monitored for cost-effective priority action.

The inflows to, and outflows from, these DMAs should be metered either continuously or on a regular basis. All data reported must be valid. All zones should also be independently isolated at the external boundaries. This ensures that when flow balance or night flow analysis is carried out, the calculated net input to the zone is occurring in that particular location and not as a result of flow into or out of another zone or sub zone.

In addition to district metering, manual and or/electronic data systems can be linked to the meters; this would integrate water accountability and water loss processes within their operations

The proposed Sarina Shire Council DMAs have been designed with respect to the initial site investigation and information supplied including the working set of paper plans.

Metering requirements

In order to assess water demand and leakage, it is essential that flow meters be correctly sized for the particular flow measurement they are monitoring. Flow-metering forms the basis of the water balance and minimum night flow investigation and will determine where the greatest losses or consumption are within the Sarina Shire Council's water distribution network.

The network has been sectorised geographically into three areas. These are:

- Sarina Township Area
- Northern Beaches Area
- Armstrong Beach Area

Sarina Township has been sub-divided into three district meter zones. These are:

- District meter zone 01
- District meter zone 02
- District meter zone 03

The existing pumped zone that feeds Jillanen and the high parts of Sarina would also be incorporated as a sub-zone of zone 03. To establish the zones, one valve has been identified needing to be closed in order to separate zones 01 and 02.

The Northern Beach area has been geographically sub-divided into five-district meter zones. These are:

- Timberlands – pressure-reduced area
- Louisa Creek
- Half Tide and Salonika Beach
- Grasstree Beach
- Campwin Beach and Sarina Beach

Armstrong Beach will remain as one district meter zone, using the existing meter arrangement at the pump station.

Each zone has been designed to minimise the number of valves to establish the sectors and also reduce the need for installing extra valves. However, it maybe necessary to repair or replace existing faulty valves in order to determine the integrity of these zones. Table CS1.3 summarises the district metering arrangement.

Meter replacements

Further to the requirements for district metering, it is recommended that loggable type meters replace the existing meters at the Armstrong Beach and Sarina Township pumped area.

Cost Summary – purchase only

The approximate cost for purchase only of the district flow meters was assessed as AUS$15 000. Installation costs were estimated to be in the region of

Table CS1.3 Metering summary.

District meter zone	Main size (mm)	Main type	Remarks
Sarina Township			
Sarina Township Zone 01 (1)	200	AC	Reservoir outlet
Sarina Township Zone 01 (2)	200	AC	Reservoir outlet
Sarina Township Zone 02	150	AC	In-line
Sarina Township Zone 03	150	AC	In-line
Sarina Township High Level	150	AC	Export from Zone 03
Armstrong Beach PS			Export from Zone 03
Northern Beaches			
Alligator Creek PS			Existing meter
Timberlands	150	upvc	Pressure control area
Louisa Creek	150	upvc	In-line
Mount Griffith	250	DICL	Reservoir inlet
Half Tide and Salonika Beach	300	DICL	Reservoir outlet
Half Tide and Salonika Beach	250	DICL	Export meter in-line
Mount Hayden	250	DICL	Reservoir inlet
Grasstree Beach	250	DICL	Reservoir outlet
Campwin and Sarina Beach	250	upvc	In-line
Armstrong Beach			
Armstrong Beach PS			Existing meter

AUS\$32 000. This initial outlay is based on mechanical flow meters; the cost for electromagnetic meters would be considerably higher. In addition to the meter purchase, there was a requirement for one non-return valve for the inlet/outlet installation arrangement at Mt Griffith Reservoir. The approximate cost for a 100 mm non-return valve is AUS\$400.

It must be noted that for district metering to proceed, the meter size for some districts may change. Once detailed investigations are carried out on each district and relevant information collated, an accurate estimation of the expected peak and low flows can be determined. This will enable the correct sizing of meters as required.

Following the implementation of the district meter areas and monitoring of minimum night flows (MNF), it was recommended that a first pass leak detection survey be carried out on those zones with excessive water losses. The cost of such surveys would be based on the total length of water mains in a particular district meter area. However, as an indication, the cost of surveying the total length of mains (130 km) in the Sarina Shire Council area would be approximately AUS\$16 000.

Action plan

Leakage data (nightlines) for all zones will need to be monitored and reported on a consistent basis to provide comparable data, trends and recorded information. A cost–benefit analysis can then be carried out on pressure reduction for individual zones with excessive pressures. These data should be used to prioritise and rank the various zones by nightline and volume loss in decreasing levels of leakage. The zones with the highest apparent leakage levels should be identified as a priority.

Having identified the zones of highest water loss, active leakage detection should begin. This would involve the following actions:

- Assess customer night use for both domestic and commercial properties within the zone. Carry out logging of sensitive commercial customers or large water users, together with a significant sample of smaller users. This is to ensure that relevant statistical database of night use values is obtained.
- Review leakage values to determine whether these are considered to be within acceptable levels for the zone, taking into account such things as size, topography and operational parameters. The zone can then be ranked according to the revised data. Should the anticipated level of leakage be unacceptable, then proactive leak detection should be carried out.
- Actively search for leakage within the system employing methods such as sampling, drop tests, step testing and correlation in an organised regulated manner to locate major levels of leakage. Carry out repairs as necessary. Reassess the zone/district on the basis of any achieved reduction. It should be noted that care must be exercised to ensure that remedial works produce a permanent rather than temporary reduction in leakage. If the reduction is permanent, revise the ranking order of the zone.
- Implement the application of leakage control through pressure management on those zones with excessive average zonal night pressures as outlined in this report. This will include calculation of:
  - reduction in background losses
  - reduction in consumption

Implementation of the action plan will:

- reduce losses
- improve system performance
- increase asset life

KEY CONCLUSIONS

1   Significant financial benefits to Sarina Shire Council in implementing a water demand management strategy as recommended in this report.

2   The demand management programme provides Sarina Shire Council with an attractive rate of return on investment and improves service delivery benefits to customers.

3   Sarina Shire Council should undertake a wide range of other WaterWise initiatives, such as school and community education programmes, shower rose rebate programmes, customer water audits and programmes focused on commercial users aimed at encouraging the local community to be more water efficient.

Case Study 2

Leakage control in southern Europe (Italy)

BACKGROUND

The Italian water industry is very fragmented, managed as it is by over 8000 different organisations ranging from public authorities to privately owned companies. The networks themselves vary enormously in size and efficiency with a large proportion run in a reactive way by the municipality inspector. The water systems often lack the necessary maintenance and rehabilitation to provide an adequate service to the customers. Mains records are either non-existent or have not been updated in years. Leakage represents over half of the total production with the worst networks approaching levels of 80% or more, due largely to the passive approach adopted to leakage control, whereby a leak is repaired only when the water becomes evident on the surface. Rationing of supply during the summer months is a frequent necessity.

The exception to such a dramatic situation is usually found in the larger cities where the networks are managed by multi-disciplinary authorities which have invested to varying degrees, in advanced management technology. Here it is normal to find good mains records, often computerised on GIS, coupled to a systematic

© 2003 IWA Publishing. *Losses in Water Distribution Networks* by Malcolm Farley and Stuart Trow. ISBN: 1 900222 11 6

approach to leakage control using the latest acoustic instruments such as correlators and ground microphones. It is still rare though, even in the most advanced organisations, to find the routine use of mathematical simulation models and active leakage control by district metering.

However desperate such a situation might at first appear, the following two case studies clearly demonstrate that through the application of advanced computerised systems coupled to the Italians' natural and unquenchable desire for innovation, a significant improvement in the efficiency of the water networks is not only feasible, but a practical proposition, even to the point where it becomes an international show case.

MUNICIPALITY OF GUBBIO

The water network of Gubbio

- *Location* (Figure CS2.1): The city of Gubbio in Umbria in central Italy.
- *Characteristics:* typical medieval Umbrian hill town which is rapidly extending to the lower-lying plain.
- *Population:* 25 000 people.
- *Type of organisation:* small rural municipality.
- *Supply sources:* bore holes pumping from the low ground to three service reservoirs located at the high point of the city.
- *Total production:* 3.1 million m³/year.
- *Storage capacity:* around 3500 m³.
- *Material:* mainly steel, ranging from DN 350 mm to DN 25 mm.
- *Pressure range:* from more than 200 m in the supply system to less than 20 m in the distribution system.
- *Original leakage level*: 60% of production.
- *Type of consumption:* mainly domestic with very little industrial use.
- *Management of the network:* three municipality plumbers
- *Main problems experienced:* low pressure in the distribution network at peak times and the need to ration supply during the summer months.
- *Quality of mains records:* non-existent.
- *Approach to leakage control:* passive.

Objectives of the project

- Reduction in the existing leakage level.
- Creation of a permanent leakage control system.

Figure CS2.1 Location map – Gubbio.

Approach

- Detailed survey of the key mains of the network.
- Construction of a mathematical simulation model.
- Field monitoring of pressure and flow at around 50 points.
- Calibration of the model to an accuracy of ±1.5 metres for three snap-shot periods.
- Design and realisation of a permanent leakage and pressure control system.

Main results

- Preparation of detailed digitised mains records.
- Reduction of leakage to below 40% with an anticipated reduction to around 30% when the replacement of 2 km of badly corroded mains is completed (Figure CS2.2).
- Application of a network analysis model for designing network extensions.
- Creation of a permanent leakage and pressure control system.

Difficulties experienced

- An initial reluctance to accept a different approach to network management.
- Lack of municipality personnel to undertake the initial survey and monitoring work due to other more urgent operational activities. This was later resolved by employing local contractors.
- Lack of appreciation of the importance of updating and maintaining the systems.

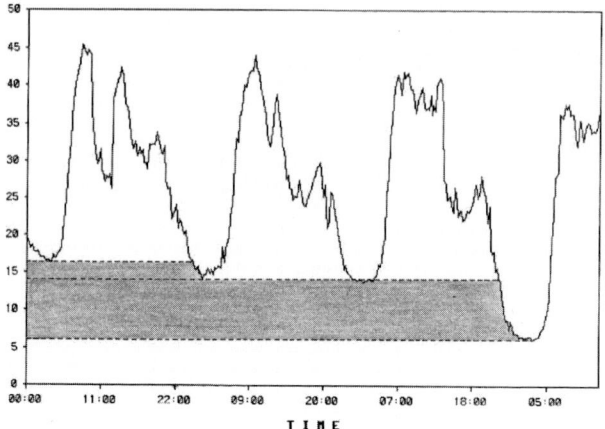

Figure CS2.2 Reduction in leakage.

Conclusion

The success of the project is best illustrated by the fact that, despite the exceptionally hot summer, 1999 proved to be the first year in a long time that water rationing in Gubbio was not necessary.

ASM BRESCIA S.p.A
The water network of Brescia

- *Location* (Figure CS2.3): The city of Brescia in the north of Italy around half way between Milan and Venice.
- *Characteristics:* An old Roman city, famous in the past for its armaments industry.
- *Population:* 200 000 people.
- *Type of organisation:* multi-disciplinary authority which in addition to the management of the water network is also responsible for the city's sewer, gas and centralised heating networks, the production and distribution of electricity, the treatment of sewage, rubbish collection, street lighting and public transport. Considered to be the most advanced authority in Italy.
- *Supply sources:* 30 bore holes pumping from the low ground to the south of the city and two spring sources to the north.
- *Total production:* 29 million m³.
- *Storage capacity:* around 30 000 m³.
- *Material:* mainly cast iron, ranging from DN 500 mm to DN 50 mm.

Figure CS2.3 Location Map – Brescia.

- *Pressure range:* from a minimum of around 20 m near the reservoirs to a maximum of 60 m in the low-lying part of the network.
- *Original leakage level:* 30% of production.
- *Type of consumption:* domestic, commercial and industrial.
- *Management of the network:* staff of over 250 people including design engineers, control-room technicians and inspectors.
- *Main problems experienced:* variability of output from the springs and bore hole which pump directly into supply, limits the scope for optimising the network without the use of mathematical simulation models; no permanent leakage control system.
- *Quality of mains records:* excellent, on a GIS.
- *Approach to leakage control:* two dedicated leakage control teams equipped with correlators and ground microphones who systematically check the network for leaks.

Objectives of the project

- Creation of an automatic system to build and maintain mathematical simulation models.
- Construction and calibration of a 24 hour dynamic simulation model of the water network of Brescia.
- Application of the calibrated model to design a permanent leakage control system for Brescia.
- Development in a pilot area of a water quality model.

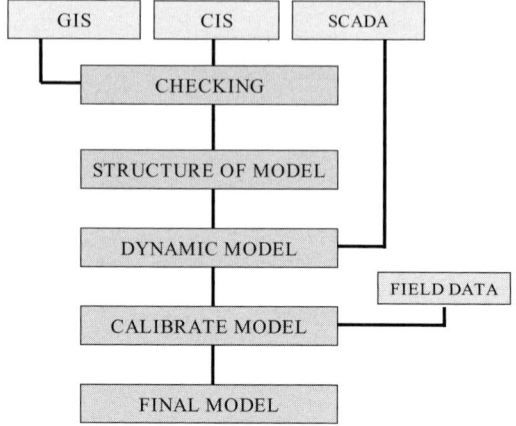

Figure CS2.4 Automatic construction process.

Main results

- Creation of a link between a mathematical model and the other existing computerised systems (mapping system GIS, billing system CIS and telemetry SCADA) to automatically construct a dynamic, all-mains model of the network of Brescia (Figure CS2.4).
- Execution of 7, week-long, field tests to measure flow and pressure at over 450 measuring points.
- Monitoring of over 200 properties to derive typical domestic, commercial and industrial demand profiles.
- Calibration of the 12 000 node model to an accuracy of ±0.5 m or better.
- Design of 35 leakage control districts.
- Construction and calibration of a water quality model which highlighted sedimentation and chlorination problems.

Difficulties experienced

- The fact that a link with a mathematical model had not been envisaged when the existing computerised systems had been specified, complicated significantly the development of the automatic model-building process.
- The resistance to the introduction of new technology, from some directors and experienced technical staff within the organisation, delayed the subsequent application of the models for network optimisation and design activities.

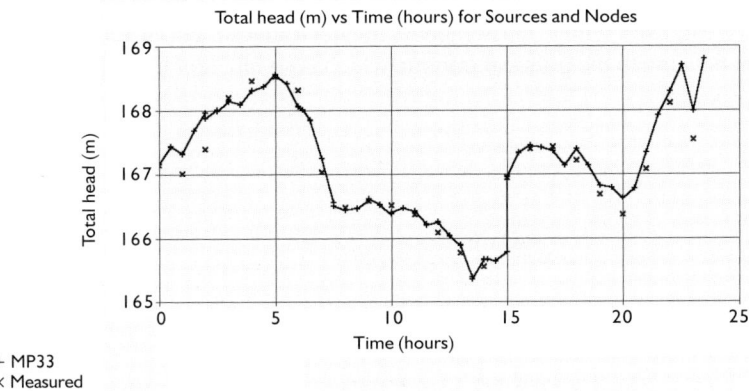

Figure CS2.5 Comparison of pressure between model and field data.

Evaluation of the success of the project

To successfully create automatically from the existing computerised archives an all mains, dynamic model of an extremely complex network which has a verified accuracy of better than ±0.5 m (Figure CS2.5), represents a significant achievement. For this reason, the project has received international recognition.

THE FUTURE IN ITALY

In an attempt to rationalise the operation of the water networks and introduce much needed investment for rehabilitation, the Italian government introduced a law in 1994, called the Galli Law, which defined an integrated approach to the management of the water cycle. After much consultation, the country was divided into around 100 areas or '*ambiti*' which represent the optimum catchment areas. The implementation of the new configuration has only just started and is likely to take many years to complete. Some of these *ambiti* will be awarded on a 20- or 30-year concessions to private companies; others will be managed by a mixed company comprising public and private participation. It is not yet clear what the procedure will be when an existing large authority is located within the *ambiti*. Most likely, local political pressures will ensure that the management of the smaller networks will be integrated slowly into the structure of the larger authority, thus avoiding the need to go out to tender.

Whatever the final configuration, it is clear that the slow process towards privatisation in Italy offers enormous investment potential for private companies. Furthermore, the projects which have been undertaken so far, clearly demonstrates that the scope for significant improvements in the efficiency of the Italian water networks is great, even with a fairly modest financial investment. It remains to be seen though, how much the political intrigue which is so inherent in Italian life will hinder the process.

Case Study 3

Leakage control and non-revenue water analysis – Water Works Corporation, Malta

BACKGROUND

This case study provides an overview of the concepts and methodologies related to the practical implementation of leakage control and non-revenue water analysis. The paper delves into the various barriers and obstacles that a water utility will encounter, with particular reference to experience gained in the Maltese Islands. The following material is covered:

- overview of the Maltese Islands; the local scenario.
- non-revenue water: concepts and terminologies.
- the first major hurdle, understanding non-revenue water.
- developing a water balance, implementation of a regional zoning scheme, obstacles to be overcome.
- real losses: implementation of the ideal methodology, restructuring, resource allocation and outsourcing.

© 2003 IWA Publishing. *Losses in Water Distribution Networks* by Malcolm Farley and Stuart Trow. ISBN: 1 900222 11 6

- apparent losses: implementation of specialised projects, utilisation of expertise from the private sector.
- sustaining the system: maintenance and database management issues.
- training and leading staff, cultural and educational issues.

OVERVIEW OF THE MALTESE ISLANDS, THE LOCAL SCENARIO

The Maltese Islands consist of three separate islands; Malta, which is approximately 95 km^2 and has around 180 000 properties, Gozo, which is approximately 26 km^2 and has around 20 000 properties and Comino, of negligible proportions. A total of nearly 370 000 habitants live on the islands, increasing to over 500 000 in the peak tourist season. The water distribution system is operated by one organisation; the government-owned Water Services Corporation (WSC), employing around 1200 full-time staff members.

All properties on the islands are metered and subsequently billed. The national water distribution network consists of around 3400 km of mixed pipe network (mainly galvanised iron, cast iron, ductile iron and polyethylene), divided into eight master zones, 40 cluster zones and 300 zones. The WSC commenced studies and research into various schemes and techniques related to non-revenue water (NRW) as long ago as 1990. In 1994 a series of initiatives were implemented which, to a large extent, changed the methodologies and practices of the WSC in relation to NRW activities.

As a result of these initiatives, national system demand reached a peak in early 1995 before starting to drop. Notwithstanding a slight increase in billed consumption over in recent years, system demand has been reduced to 33% below its 1995 values, to the extent that the 1999 system demand equalled the 1989 system demand (Figure CS3.1). Calculations indicate that national leakage has been halved, with leakage figures comparing favourably to international values. Optimum economic leakage targets have been set and, in some localities, achieved.

NON-REVENUE WATER CONCEPTS AND TERMINOLOGIES

'Non-revenue water' (NRW) refers to an accumulated range of losses experienced by a water utility when comparing the network system demand with the quantity of water that is known to used by network consumers. The components are illustrated in Figure CS3.2. Throughout this case study, the following terminology is used:

- System demand: the quantity of water input into the water network, often also referred to as *system input* or *production less reserves*. System demand can be

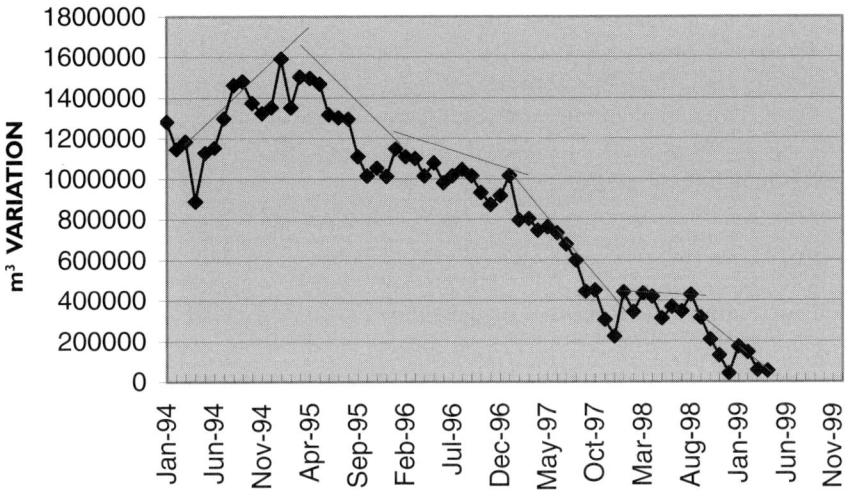

Figure CS3.1 National system demand compared to 1989 baseline.

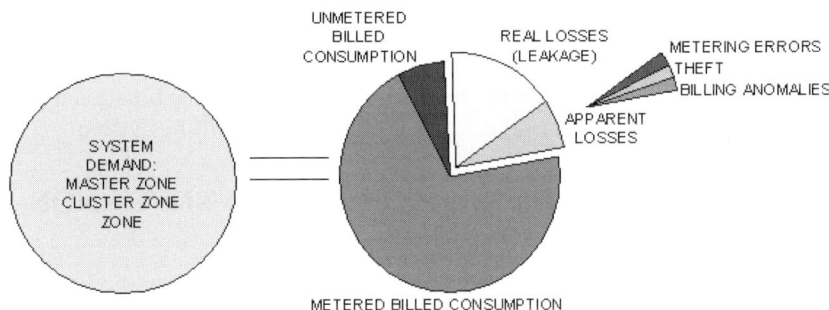

Figure CS3.2 Non-revenue water components.

computed for any metered and hydraulically encapsulated water networks, such as master zones, cluster zones or zones.
- Hydraulically encapsulated area: a part of the water distribution network that is separated from the remainder of the system by shut valves, capped mains and a minimum number of metered interaction points.
- Master zone: a large hydraulically encapsulated segment of the water utility's distribution network, comprising numerous zones, reservoirs, production sources, trunk/transfer mains, etc.

- Cluster zone: a smaller, more manageable part of a master zone that is also hydraulically encapsulated and usually comprises a limited number of zones, reservoirs and production sources.
- Zone: a segment of the network, hydraulically encapsulated, preferably with just one metered (and data-logged) supply inlet, the zone size being ideally fewer than 3000 properties.
- Real losses: this first component of NRW consists of all forms of leakage within the network, such as service pipe leakage, leakage on fittings, reservoirs, trunk/transfer/street mains, etc. Any leakage downstream from a production source and upstream of the consumer revenue meter is termed a *real loss*.
- Apparent losses: sums three principal sources of NRW that result in water that is actually consumed (or utilised), but not successfully billed. For this reason these three components are not a *'real loss'* (as is leakage) but an *'apparent loss'*.
  - Metering errors: this first component of the apparent loss range can be subdivided into two categories: a) revenue meter under-registration, resulting in a lower than actual calculation of consumer water usage and b) production meter over-registration, resulting in a higher than actual calculation of system demand.
  - Water theft: this second component of the apparent loss range consists of the illegal or unauthorised usage of water taken from the system.
  - Billing anomalies: this third component of the apparent loss range includes a multitude of factors that contribute to a distorted picture of legitimate consumer usage due to the ineffectiveness of the water utility's billing system.

THE FIRST MAJOR HURDLE: UNDERSTANDING NON-REVENUE WATER

NRW is, unfortunately, often quantified in percentage terms, using the following equation to depict the cyclic or annual NRW for a water utility:

$$NRW = \frac{System\ demand - System\ consumption}{System\ demand} \times 100\%$$

This NRW ratio, whilst being simple to compute and easy to understand, often gives a distorted picture of the true NRW content for a water utility since it depends upon the volume of billed consumption. Hence, for example, two utilities with identical real and apparent loss components will have entirely different NRW percentage values if the average per capita consumption of one utility is different from the other.

Any water utility that plans to substantially improve NRW analysis and control has to overcome this first major hurdle. Senior corporate management, company stakeholders and shareholders, and consumers, all have to be educated sufficiently so that NRW can be properly quantified and understood in relation to internationally recognised benchmarks. Failure to achieve this may result in setting NRW target levels which are either impossible to achieve or are not economically viable.

Table CS3.1 clarifies this issue. This table, issued by International Water Data Comparisons Ltd., in association with the International Water Services Association, shows Gozo as reference number 3. The table compares various benchmarks used to gauge real losses with a calculation of real losses in percentage terms. In row 3, Gozo is third in the set of 27 in terms of ILI and leakage computed in litres/propery/ day, but sixteenth in terms of percentage leakage due to the low per-capita consumption. This master zone, with 15.8% leakage, actually has a leakage level lower than the optimum calculated economic leakage level.

Whilst this example shows only part of the NRW content – real losses – the same argument holds true for apparent NRW components. Benchmarks such as the simple m³/hour, l/prop/day or m³/km/day should be used whenever and wherever possible to compute any of the three major components of apparent losses. Realistic target setting can now be put into play.

DEVELOPING A WATER BALANCE, IMPLEMENTATION OF A REGIONAL ZONING SCHEME, OBSTACLES TO BE OVERCOME

The most critical task for any NRW management team is to develop a methodology whereby the quantification of the four NRW factors can be made. In other words 'if you can measure it … you can manage it …'.

In this regard the primary objective of any water utility is to implement a zoning scheme whereby the complete water distribution network is broken down into manageable segments that can be easily metered, monitored and analysed. The advent of telemetry or SCADA (supervision, control and data acquisition) systems may simplify the data acquisition and control processes but it must be appreciated that the moving from more simplistic data logging to various levels of automation only makes economic sense once the zoning scheme has been successfully and fully implemented.

The implementation of different zoning levels or hierarchies facilitates the development of a 'source to sink' mode of comparison of flows within the network. It is this flow analysis, shown in simplified form in Figure CS3.3, that ultimately leads to the building of an effective water balance. The implementation and sustaining of a proper zoning scheme is hence the groundwork for all further operations related to NRW analysis and control.

Table CS3.1 World-wide international leakage indices compared with alternative performance indicators as at 23 January 1999.

Supply system ref no	Average pressure (m)	Density of conns (conn/km)	Average length of pipe from street (m)	International leakage index (ILI)	Real losses (l/conn/d)		Consumption (l/conn/day)	Real losses % of system input	
						rank			rank
1	35	55	3	0.45	19	1	539	3.5	3
2	50	53	3	0.53	32	2	525	6.1	6
3	45	103	0	0.97	42	3	266	15.8	16
4	40	38	10	1.21	74	5	7,824	0.9	1
5	57	47	30	1.31	146	12	496	29.4	23
6	106	28	0	1.32	202	14	1,023	19.7	19
7	35	39	6	1.55	76	6	1,734	4.4	4
8	46	72	14	1.62	104	7	533	19.5	18
9	60	55	0	1.66	114	8	1,199	9.5	8
10	46	71	23	1.72	130	9	543	23.9	21
11	39	86	0	1.77	70	4	1,280	5.5	5
12	57	45	0	1.94	132	10	1,142	11.6	12
13	54	48	20	2.91	263	16	511	51.5	25
14	30	35	10	2.94	138	11	5,633	2.4	2
15	70	31	5	3.25	342	18	4,230	8.1	7
16	30	65	5	4.99	180	13	567	31.7	24
17	46	37	0	5.42	320	17	2,208	14.5	14
18	50	58	5	5.94	367	19	2,552	14.4	13
19	36	29	15	6.21	401	21	719	55.8	26
20	39	26	10	6.44	436	22	3,004	14.5	15
21	35	29	20	7.06	477	23	2,652	18	17
22	31	79	0	8.02	256	15	2,481	10.3	9
23	48	114	0	8.04	370	20	1,742	21.2	20
24	71	21	0	8.15	956	27	1,669	57.3	27
25	45	24	10	9.39	759	25	6,921	11	10
26	37	27	10	9.48	600	24	5,214	11.5	11
27	45	24	10	10.25	832	26	3,230	25.8	22

This table contains data provided by water undertakings in Brazil, Denmark, France, Finland, Germany, Gibraltar, Greece, Iceland, Japan, Gozo, Netherlands, New Zealand, Singapore, Spain, Switzerland, Sweden, UK, USA and West Bank (Palestine) for various 12 month periods between 1995 and 1998. International Water Data Comparisons Ltd.

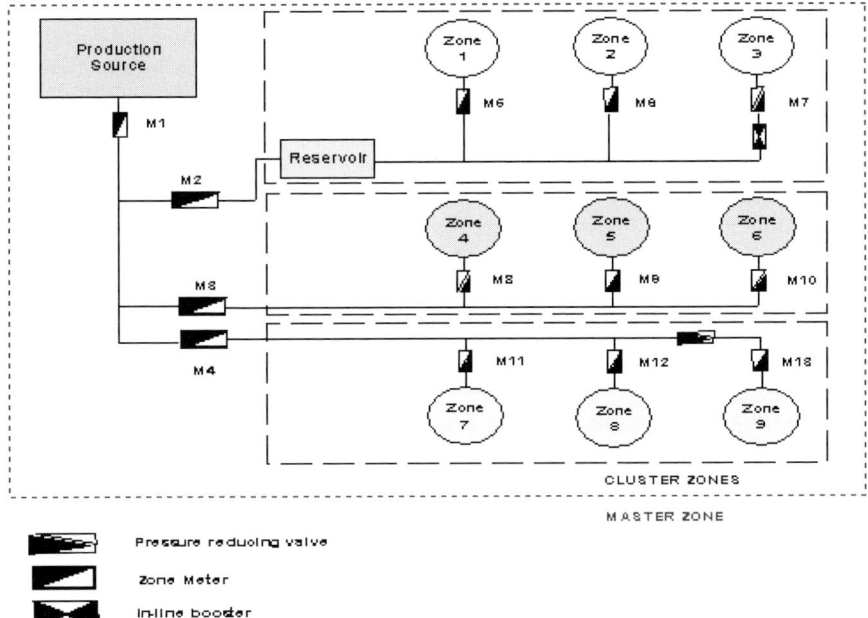

Figure CS3.3 Zone hierarchy.

PRACTICAL IMPLICATIONS IN THE IMPLEMENTATION OF A NATIONAL ZONING SCHEME

As was the case in Malta, many networks develop gradually over time, with little consideration to zoning criteria. It usually takes a number of years for the successful implementation of a national zoning scheme as a prelude to the creation of a water balance. The following issues usually have to be resolved during the course of this development:

1. **Corporate commitment:** at this early stage of investment and development, corporate commitment is a major issue. The corporation must understand the *vision* of what is going to be achieved and be ready to commit the necessary resources, without having the benefit of exact costings and targets.

2. **The issue of short-term vs. long term goals:** even in the early stages of development, zone information can be utilised wisely to target high leakage areas for leakage control purposes. Hence, short-term savings can be achieved from the start, although the long-term target may take years to complete.

3. **Equipment and instrumentation:** various new systems and techniques have to be tried, implemented and ultimately sustained, that are compatible with the

local environment. In the Malta scenario, these included helical vane zone meters, flow/pressure data loggers, pressure controllers, acoustic and correlation detection equipment, ultrasonic clamp-on meters, digital pipe-locators, etc. All the systems had to be chosen to be appropriate for the potential level of expertise of local staff.

4. **Choice of methodologies:** part of the necessary research and development also includes testing different methodologies together with the choice of instrumentation. This, and further points below, are described in greater detail when discussing real and apparent losses.

5. **Staffing and structuring:** again, in conjunction with the above issues, restructuring is usually inevitable at an early stage in the development period. Specialised and dedicated ongoing training of the different levels of staff is an integral part of this staffing process.

6. **Data and data management:** this will include aspects from as simple leakage database creation to more advanced GIS development.

7. **Corporate culture change:** the shift from the concept of simply supplying water at adequate pressure to creating and sustaining a low leakage level system demanded a slow but important cultural change in the attitude of existing staff.

8. **Consumer awareness and education:** implementation of a national zoning scheme will unavoidably cause disruptions and shut-offs for consumers. This is the best time to implement a PR campaign aimed at increasing consumer awareness and appreciation of the aims of the programme.

9. **Outsourcing of work:** contracting out of excess workloads and certain specialised tasks may be an ideal way of balancing existing resources with the fluctuating amount of work required.

The remainder of this case study will now focus more closely on the individual parameters of the NRW equation.

REAL LOSSES: IMPLEMENTATION OF THE IDEAL METHODOLOGY, RESTRUCTURING, RESOURCE ALLOCATION AND OUTSOURCING

A well-balanced approach towards active leakage control demands the correct methodology and the implementation of related staffing and resource allocation. This was made clear to the WSC in the years leading up to the 1995 peak system demands, where it became evident that the existing 30-person team haphazardly sounding the water network approximately 3 times yearly, was far from adequate.

Figure CS3.4 indicates the four factors that constitute an ideal leakage control methodology. Failure of any one of these factors usually results in an inability to

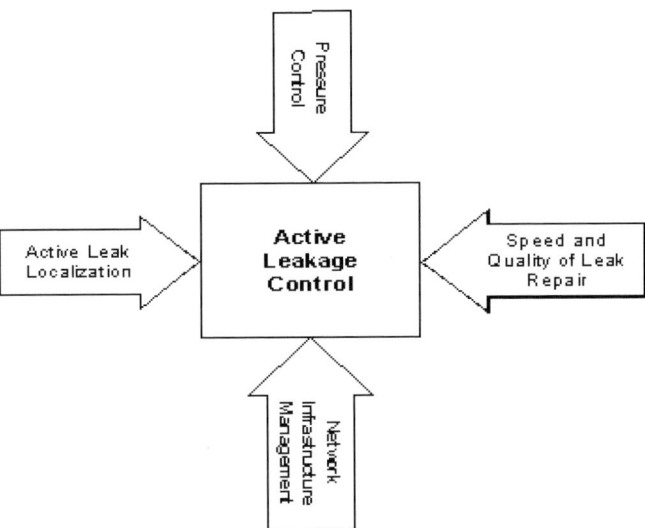

Figure CS3.4. The four pillars of active leakage control.

achieve significantly low leakage levels, or a failure to sustain these levels once they are achieved. In additionon, an understanding of the economics of leakage control and the computation of optimum economic leakage levels is also essential.

Throughout the development and implementation of the chosen leakage control methodology, a number of key issues usually need to be resolved. These are:

- **Research and development:** various leakage detection techniques are usually trialled such as acoustic detection, correlation techniques, infra-red scanning and leak noise data logging. A range of pressure and flow data loggers and their interface with existing and available distribution water meters, should also be assessed.
- **Specific environmental constraints:** The effect of soil and backfill conditions, network materials, ambient and background noise, existence of other utility's underground pipe, cable, gas systems, etc. all should be assessed and understood.
- **Accountability of staff through proper structuring and target setting:** in a task as intangible as leakage control, the concept of setting accountability levels through an adequate, functional and decentralised structure is imperative. The structure shown in Figure CS3.5 was successfully implemented by the WSC, directly addressing all areas of the NRW range.
- **Economic leakage level target setting:** calculations must be made to ensure that the economic leakage levels for each master zone are computed, understood and targeted.

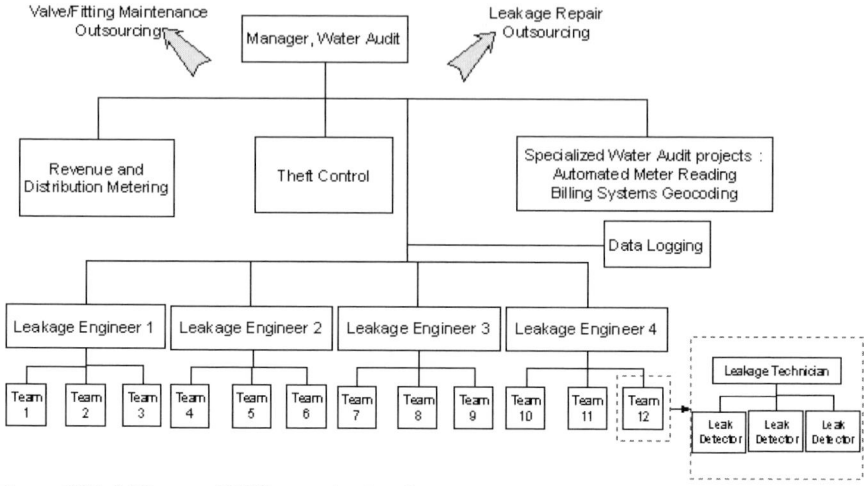

Figure CS3.5 The new WSC organisational structure.

- **Contracting out of excess workload:** the outsourcing of certain tasks helps reduce the problems caused by a continually fluctuating workload related to leak control. As shown in Figure CS3.5, the WSC decided to contract out two major tasks: system maintenance, and leakage excavation and repair. Coordination and communication between full-time staff and contract staff is often a delicate issue.
- **Union and institutional constraints:** No water authority functions in isolation. A powerful trades union, for example, can hamper plans to implement what management may see as an ideal leakage shift layout. The leakage manager often performs a balancing act in his efforts to push his concepts through.
- **Day–night shift operations:** the ideal working scheme for the chosen methodology must be implemented. This usually involves substantially night duty and a certain degree of flexibility to allow the leakage teams to work both day and nights as necessary. In the case of the WSC, the concept of flexi-time was introduced for the first time.
- **Feedback and control:** again, due to the intangible nature of leakage control, continuous and exhaustive evaluation of achievements and results is required.

APPARENT LOSSES: IMPLEMENTATION OF SPECIALISED PROJECTS, USE OF EXPERTISE FROM THE PRIVATE SECTOR

The complex issue of quantifying and reducing the various components of apparent losses often involves the implementation of dedicated projects on completion of research in a particular field. The following are examples:

Metering schemes

A water authority must carry out a study into the performance of different types and classes of revenue meters appropriate for domestic consumption trends. Various trials involving high frequency data logging of statistically selected consumers and the testing of a range of meters on the achieved consumption trends, are imperative.

In the Malta scenario, site and laboratory research led to the selection of the class D (Q_n=1.0 m³/hr) volumetric meter model. The task of replacing over 100 000 old class C meters with this model was subsequently contracted out to six chosen private companies.

Theft control

This delicate issue involves institutional aspects such as a detailed study of legal and criminal codes of practice and also the issue of relations between the water utility and other organisations. Furthermore, proficient water theft control teams have to be trained, put into operation and carefully supervised. Advanced pipe location equipment is often a necessary asset.

Study of the Maltese criminal code and discussions with leading legal experts were the first steps taken locally. Minor amendments to the legal framework were made before the theft control teams started operation. Steps are now being taken to contract out water theft control.

Billing aspects

A number of possible complications and discrepancies caused by flaws or inefficiencies in the water authority's billing system exist. An example of this is the failure of an authority to read consumer meters installed inside properties, resulting in 'guessed' estimations. Two vital projects can help quantify and subsequently reduce this type of *apparent* loss:

- **Geographical information systems (GIS):** A GIS will allow the water authority to map consumers onto '*points*' on a map layer. Dedicated software can then compute the total billed consumption for each zone. Since the system demand and the leakage level of the zone are already known, the total apparent loss value can be computed.

- **Automated meter reading (AMR):** Use of AMR can reduce substantially the apparent losses related to billing, thus allowing for the calculation of apparent losses due to revenue meter under-registration. Expert help in this field can be found through organisations such as the Automated Meter Reading Association.

SUSTAINING THE SYSTEM: MAINTENANCE AND DATABASE MANAGEMENT ISSUES

Sustaining the methodologies that have been implemented is often not an easy task. All resources have a finite life and must be regularly maintained, calibrated, replaced, etc. Problems may arise with regard to standardisation or purchasing issues. The more an organisation diversifies in its purchasing and maintenance schemes, the more it becomes open to long-term complications. Use of databases, such as the national leakage database depicted in Figure CS3.6, can help substantially in managing resources.

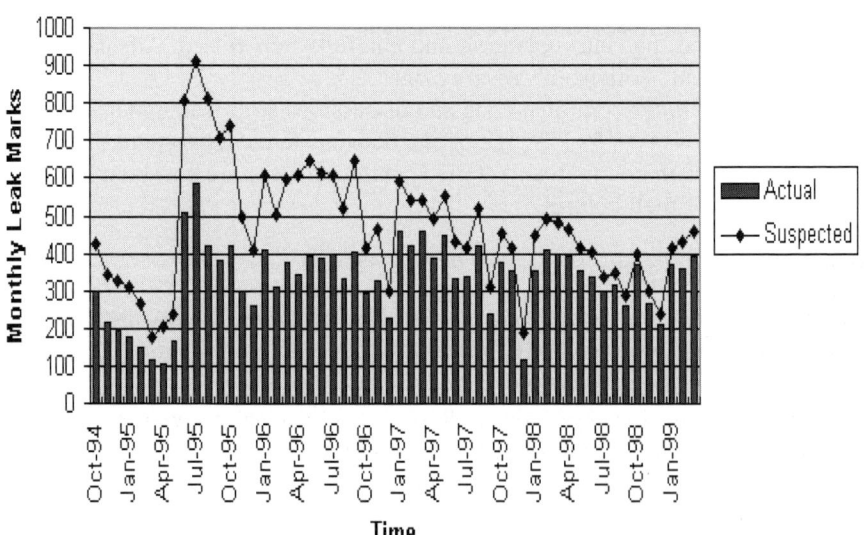

Figure CS3.6 National leakage database.

TRAINING AND LEADING STAFF, CULTURAL AND EDUCATIONAL ISSUES

The management of NRW control involves a high degree of leadership and empowered teamwork. The leader must also be a team member and must involve himself in the daily difficulties and challenges faced by his workforce if he wishes to maintain their respect and esteem. The implementation of NRW methodologies often results in a culture clash where 'old' staff are required to change their mentality and attitudes to work and responsibilties. Where, for example, an underground leakage may have previously been hidden and a problem to nobody, now the team has become empowered to quantify and locate it, and is being held accountable for its elimination.

As was the case in Malta, water authorities may be faced with the sometimes daunting task of having to implement advanced techniques using staff with low educational levels. Ongoing training programmes set up through a permanent training centre can target most education requirements except the most advanced, for which outsourcing of training is a solution.

Bibliography

Lambert, A, Myers, S and Trow, S (1998) Managing Water Leakage – Economic and Technical Issues. London: Financial Times Energy.

Rizzo, A (1997a) *Report on the Development of Leakage Control and Water Auditing in Gozo.* Valetta, Malta: Water Services Corporation.

Rizzo, A (1997b) *UFW Analysis.* Engineering Today (Chamber of Engineers) Malta: Crest Publicity Ltd.

Rizzo, A (1998) *A Review of Water Accounting in Malta.* Valetta, Malta: Water Services Corporation.

Water Services Corporation (1997) *Annual Report.* Valetta, Malta: Water Services Corporation.

Water Services Corporation (1998) *Annual Report.* Valetta, Malta: Water Services Corporation.

WSA/WCA Engineering and Operations Committee (1994) *Managing Leakage: UK Water Industry Managing Leakage* Reports A–J: Report A – *Summary Report*; Report B - *Reporting Comparative Leakage Performance*; Report C – *Setting EconomicLeakage Targets*; Report D – *Estimating Unmeasured Water Delivered*; Report E – *Interpreting Measured Night Flows*; Report F – *Using Night Flow Data*; Report G - *Managing Water Pressure*; Report H – *Dealing With Customers' Leakage*; Report J – *Techniques,Technology and Training.* London: WRc/WSA/WCA.

Appendix A

Potential savings from leakage management

Cost savings from leakage management come from a variety of sources. There will be savings in ongoing operating costs, and savings from capital works investment costs. Savings will be associated with the water distribution system, where much of the leakage will be saved, but they will also be linked to the water resources and treatment works which supply the area where leakage has been reduced.

DISTRIBUTION SYSTEM SAVINGS

Reduction in network flows due to lower leakage levels

Lower leakage in an area may result in reduced power costs for boosting and pumping the water around the network.

Reduction in burst main frequency

Mains replacement and pressure management to reduce leakage will have the additional benefit of a lower future burst frequency resulting in lower cost of repairs, and a lower cost of dealing with the impact of these events on the network and on customer supplies.

Once a lower mains and service pipe failure level has been established, it will also be possible to reduce the level of emergency work and hence the number of staff on standby or call-out arrangements.

Reduced customer complaints

A lower level of complaints from customers may result from the leakage management programme (although there may be an increased level while the work is in progress). A reduction in customer contact will result in lower costs of dealing with the issues which arise. Lower complaints may result from:

- reduced demand on the network leading to reduced head losses, and also due to better management of pressures, causing fewer complaints of excess or inadequate pressure at customer premises
- reduced number of bursts
- improved water quality. Correct DMA design improves water quality – as it provides a specific route for water to flow from source to customer's tap, rather than washing about in the system. It also means that network modelling can be carried out to investigate retention times in the system, and remedial action can be taken in areas of high retention times. Reduction in leakage combined with downsizing of the network, allows retention times to be reduced.

Improved co-ordination of works

Reduced overall workload will result from better coordination of leakage repairs with mains replacement activity. It may be better to replace mains which have a history of bursts. It may also be better to wait for leakage detection to be carried out after mains replacement in order to avoid repairs on mains due to be replaced

Reduced network capacity requirement

If leakage management is coordinated with mains rehabilitation planning, it will be possible to make savings due to reduced network flows allowing:

- downsizing of mains by slip lining or pipe bursting
- abandoning some mains completely as they are no longer required to meet lower demand conditions

A reduction in service reservoir storage capacity, and a reduction in boosting and pumping plant capacity, may also be possible in system with lower leakage levels.

SOURCE AND TREATMENT WORKS SAVINGS

Reduction in distribution input leading to reduced source operating costs. Lower throughput at treatment works will result in operational cost savings due to:

- power – lower electricity charges in the treatment plant
- chemicals – lower chemical treatment costs to clean and disinfect the water supply
- sludge disposal – a reduction in sludge volume requiring disposal – a significant factor in the UK, where landfill tax is levied.
  Depending on the nature of the treatment, some of these costs may be minor.

Reduction in other demand related charges

In some cases business rates and government taxes are related to the output of individual treatment works, or to the total volume of water supplied.

Reduction in long-term resource needs

If lower leakage levels can be sustained, then there will be a lower requirement for treatment and source works capacity in any future year. This will allow the deferment of new capital works to meet future demands or expenditure need to meet enhanced water quality directives/requirements.

Reduced treatment works capacity may allow the opportunity to defer more expensive sources. In some cases sources can be down-graded to seasonal use only.

Appendix B

Meter installation design and schedule of materials for a zone/DMA meter

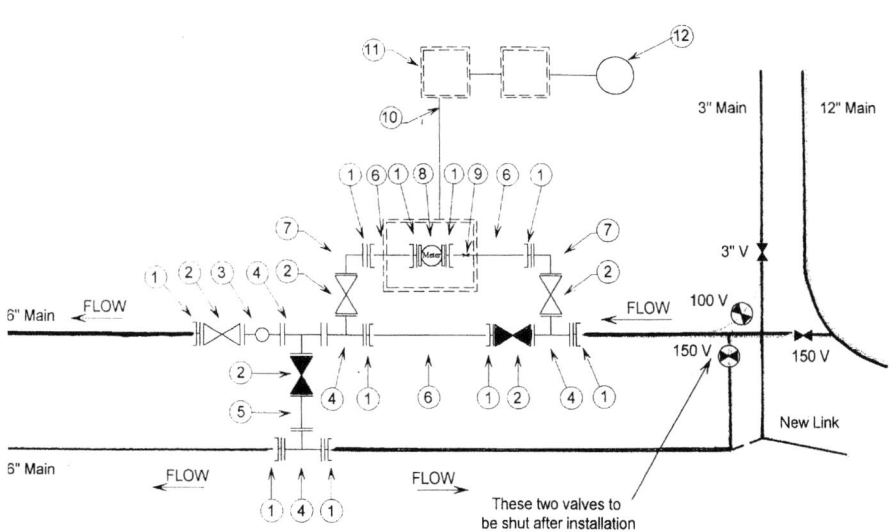

Figure B.1 Example installation diagram for a 100/150mm meter in a 150mm main.

Figure B.2   Schedule of materials for installing a 100mm/150mm meter and bypass on a 150mm main

Ref. No (see diagram)	Materials	Quantity
1	150 mm flange adapter	10
2	150 mm SV	4
3	150 × 150 × 80 mm tee with W/O	1
4	150 × 150 × 150 mm tee	2
5	150 ductile iron pipe – length to suit	3
6	150 mm ductile iron pipe – length 0.75m	1
7	150 mm 90° elbow	2
8	150 mm meter	1
9	Pressure tapping	1
10	150 mm ductile iron pipe – length to 0.75m	1
11	50 mm ducting	As required
12	Boxes	2
	Antenna	1
	Grade A chamber cover	1
	Communications	As required
	100 mm meter *	1
	100mm DI pipe – length to suit	As required
	150mm to 100mm tapers *	2

* alternative materials for installation of a 100mm meter

Appendix C

Network records

The following sections detail the records required for each level of the network hierarchy.

SUPPLY ZONES

- Supply zone boundary
- Zone boundary valves
- Type of source (reservoir, well, pumping station etc.)
- Supply (production) meter
- Zonal monitoring points (pressure)
- Temporary insertion meter points (e.g. reservoir outlet)

All other network features:

- transmission mains (length, diameter, material, air valves, other features etc)
- distribution network mains (length, diameter, material, valves, hydrants, other features etc.)
- customer service pipes (length, diameter, material, connection point, stop tap etc.).

© 2003 IWA Publishing. *Losses in Water Distribution Networks* by Malcolm Farley and Stuart Trow. ISBN: 1 900222 11 6

DISTRICT METER AREAS (DMAS) AND PRESSURE MANAGEMENT AREAS (PMAS)

These are 'static' records relating to:

- DMA identifier
- DMA boundary
- DMA boundary valves
- Boundary valve configuration (valve or valve + wash-out)
- DMA property numbers and classification
- DMA meter(s)
  - meter ID and type
  - direction of flow
  - strainer (where fitted)
  - maintenance/repair history
  - telemetry housing/data logger chamber
- Insertion meter tapping (insertion) points
- Pressure-reducing valves (PRVs)
- Pressure management area (PMA) boundary valves – if different to DMA
- Pressure monitoring points – average zone pressure (AZP) point and critical points
- Sensitive areas/customers (hospitals, dialysis patients etc.)
- Customer complaints (pressure and quality)
- All other network features as for supply zones

A DMA numbering system should follow a logical pattern. There are a number of options, and each company has its own system. One system is for the DMA to be linked by number or name to its 'parent' water-into-supply zone, which may contain several DMAs. The supply zone prefix could be followed by the suffix 'D' (for DMA) and the DMA number. Sub-areas within the DMA could be similarly linked to the 'parent' DMA.

Plans and records which may assist the leakage team are:

- Regional schematic
- Area schematic
- DMA plans
- Meter records

The preparation of plans is subject to a company's information system (in-house or bought-in, PC-based or paper-based). Some companies use GIS to generate plans, mounted on portable PCs for field use. The PCs can also contain software for entering DMA data and the algorithms for calculating net night flows and leakage.

Regional schematic

This is a small scale (e.g. 1:10 000 or 1:25 000) map of the distribution network. It should show the DMA boundaries and metered feeds in relation to the key elements of the water into supply zones – trunk mains, service reservoirs, pumping stations, supply and pressure zone boundaries. Details of the distribution system within DMAs can generally be omitted unless there are key mains passing through one district to another.

Area schematics

Supply zone drawings can be used to produce a set of DMA plans, showing all DMAs within each water into supply zone. They will show the position of each DMA in the zone, with boundary valve and meter identifiers.

DMA plans

These are detailed plans of each DMA, identifying:

- numbered boundary valves and meters
- valves for DMA sub-division or step-testing
- large metered customers (logged)
- features of the network (mains, line valves, hydrants)
- special needs customers

Meter records

These records contain information needed for data interpretation and meter maintenance. It is recommended that the following information is recorded on a suitable form or PC file.

- DMA identifier
- meters in and out of the DMA
- meter identification or serial number
- meter type and size
- bidirectional flow
- main size
- maintenance and repair log

Leakage analysis records

Leakage monitoring and analysis is described fully in Chapters 4 and 6. Records required are those which relate to the calculation of net night flow and leakage from total night flow in each DMA, using the basic formula:

$$\text{Leakage (total night flow losses)} = \frac{\text{minimum night flow - customer use}}{\text{number of properties}}$$

$$or \quad \frac{\text{minimum night flow - customer use}}{\text{length of main}}$$

Records required are:

- night flows at each meter
- non-metered household count
- occupancy rate
- numbers of metered users in each category
- large industrial users
- allowances for night use
- net night flow
- average zone night pressure
- pressure profile at mid-zone
- hour to day factor

Night flows

These should be recorded at selected intervals over a selected period, summing and/or subtracting flows from multiple DMA meters.

Non-metered household count

This should be entered in the file for each DMA, and updated regularly.

Metered users

Record the number of metered customers in each category, preferably using Standard Industrial Classification (SIC) codes Record the total estimated night use of each category, based on customer demand studies.

Include non-metered commercial properties in domestic property count, unless they are significant night users.

Customer night use

These records relate to each of the categories of customer and their night use. Night use is an important record as it is subtracted from night flow delivered to derive leakage on non-metered household service pipes, and their plumbing losses, in the DMA.

Large metered customers

Customers who use significant amounts of water at night should have night flows recorded simultaneously to DMA monitoring.

Other metered customers

Consider a study of each category to provide a better estimate of night consumption.

Operational use

This includes water abstracted for night mains flushing and fire fighting, which, although comparatively small, may influence minimum night flow readings on a particular night.

Leak detection and localising areas

This is the sub-division of DMAs for leak detection operations – 'step-testing' or noise logging:

- Area identifier
- Step-test area boundary
- Valves to be closed (temporarily) during step-test exercise
- Correlator survey plans (sounding points and pipe lengths)
- Noise logger survey plans (clusters of hydrants)

Leak location records

Most companies use two databases, one for DMA leakage analysis, to monitor total leakage, the other a customer database, to monitor progress with service pipe repairs.

Records are needed to:

- monitor when boundary valves are open or closed
- assist the inspectors in planning a location exercise following DMA monitoring and/or a detection exercise
- direct the repair gang in finding the location of a leak
- monitor progress on repairs
- provide leak and burst records for the DMA file

The time taken for leak location and repair in each DMA should also be recorded for future analysis and decision making to assess improvements to leakage levels.

It is good practice to record the approximate leak positions on a DMA map. This builds up a picture of the leakage characteristics, and helps to identify clusters of leaks which may relate to the age of the pipe or the pipe material. This will provide useful information for a future rehabilitation strategy, and will enhance the DMA database by identifying characteristics which may affect future decisions on leak location.

Leak repair records

Following leak repair, records should be kept of:

- date of repair
- exact position of leak (if different from the estimate)
- cause and type of leak, and repair carried out
- pipe material and size, and whether pipe replacement was necessary

These records are essential for identifying weak spots in the network and assist mains renovation/replacement planning.

Appendix D

Conducting a water use audit

THE AUDIT PROCEDURE

The stages are:

- Initial discussions
- Planning
- Investigations
- Reporting and recommendations
- Staff training and awareness
- Installation of devices
- Follow-up and monitoring

Initial discussions.

There should be a pre-arranged meeting with senior staff, who should be decision makers and enablers. The purpose of the meeting is to allow decision makers to:

- become aware of the aims and objectives of the audit
- allocate a 'task force' of key personnel to assist the audit team
- give an overview of the main features of the site (process type, site geography etc.)
- give their support and commitment for authorising changes
- identify a budget for new installations

Planning

During this stage meetings will be held with members of the site task force, who should be representative of the site components and practices (e.g. office manager, technical services manager, production manager). Discussions will centre on:

- details of site geography, processes, component units (eg offices, restaurant, laundry, production units)
- plans of the layout of the site, details of the pipe network, existing flowmeters (water company meter and any sub-meters), position of line valves, pipe materials and fittings, and their general condition, replacement and repair history, capped ends etc.
- identification of suitable points to install sub-meters if necessary, and gas injection points for leak detection
- alternative water sources or supplies
- staffing organisation (who is responsible for what) their knowledge of water use and consumption (anomalies, anecdotal evidence, trouble spots, known high consumption or leakage)
- past records of metered consumption and water bills

From these meetings a plan will be drawn up for conducting the audit.

Investigations

The starting point should be a record of site consumption. This will be an invaluable demonstration of savings being made, immediately and in the future, as best practice is introduced, leaks repaired and water saving devices fitted. A consumption record is most easily achieved by installing a data logger on the main meter to the site. This meter is owned by the local water company, who should give permission, but may insist that their staff do it. First check the meter installation:

- is it working?
- is it installed correctly?
- is it the right size for the flow range?
- has the process or consumption pattern changed since the meter was installed?

It would be an advantage at the same time to measure pressure at this point. This will show whether pressure is unnecessarily high and can be reduced (reducing the flow rate and the incidence of bursts). Input pressure may also be compared with other parts of the site to check anomalies (e.g. bursts) downstream.

The 24-hour consumption pattern is then analysed. The night flow is particularly important. Questions to ask are:

- is the consumption pattern in line with what is expected from the process?
- does the process continue at night, or should consumption be the same as a domestic property?
- are there tanks filling overnight, urinals flushing, showers being used?
- can the process be temporarily stopped or re-zoned to identify a baseline for night consumption (i.e. leakage or other unexpected or illegal use)?
- are there meters downstream of the input meter? If not can they be easily installed?

The next stage is to break down the site into individual locations, or easily manageable areas. If these areas can be isolated and individually metered it is an advantage, otherwise a visual inspection of consumption pattern, water using practices and appliances, and production processes etc. will suffice. The range of use is huge, and is site dependent, but look for excessive use. Examples are:

Industrial processes

Initial discussions will help to clarify the process stages and any water used in each. Use specialised client knowledge to investigate potential for changes to the process, e.g. can process water be recycled? Is there an alternative source on the site – private borehole or well, or can non-potable water be used? This may require investment in a small water treatment plant, and a separate cost benefit study would be needed.

Production units

Look for staff facilities, such as washing, showering, and laundry, as well as the process itself. Is there scope for flow restricting devices?

Office units

Schools, public buildings etc: minimal use at night (can be compared with national household night consumption figures). Look for savings on urinal flushing (sensing devices, timed stop valves). Consider water use for landscaping and gardens – are sprinklers being left on overnight?

High use units

Hospitals and retirement homes have a high night use. UK national figures exist for expected use for these categories (Managing Leakage, Report E [1]).

The analysis of these data will show if night consumption is as expected. If consumption is high for no apparent reason, there is probably one or more bursts or smaller leaks in the system. Some of these may be picked up from a visual inspection, but most will need a leak location exercise. Depending on the site this can be done by:

- gas injection (identify suitable injection points)
- leak noise correlator (identify suitable sounding points)
- ground microphone

Each has advantages in different site situations. Gas injection is best on indoor sites, under concrete etc., when precise location is needed, and for small diameter service pipes. The correlator is best for long sections of distribution main, where the precise measurements required for its accuracy are more easily made. The ground microphone could be used as a cheaper (but less accurate) alternative to the correlator, perhaps for bursts on exterior sections of pipeline. These instruments have been described in Chapter 6.

The investigation stage will also include noting the scope for installing water saving devices, and pressure reducing valves etc. These should be costed, with pay-back benefits calculated from estimated savings on the water bill, and included in the report

Reporting and recommendations

The report should contain:

- a description of the audit, indicating the stepwise approach taken
- any anomalies found from the meter audit
- a record of 24-hour night consumption from the data logger (and any subsequent reduction in consumption arising from changing practices etc.)
- results of investigations and findings, broken down into geographical areas where feasible
- recommendations for changes in practice, process, and water use habits etc.
- recommendations for installation of water saving devices, with an estimate of installation costs and payback.
- recommendations for staff awareness training, and follow-up checks
- a suggested process for monitoring future consumption

Staff training and awareness

As with all awareness programmes, presentation of the findings, based on the report, starts with senior staff and decision makers. This can take the form of a short awareness seminar to take them through the findings and the cost benefits of changes. It will reinforce their commitment, and give them an opportunity to allocate a budget. Ideally, the recommended changes will be built in to a management system. In subsequent seminars, other staff will be made aware of the audit findings. They will hear about any changes made to the process, and will require awareness training for using water saving devices, conservation techniques, and any new practices.

Installation

Some improvements to the current practice and procedures are no cost or low cost to the client and can begin immediately during the audit, as anomalies are highlighted. As such improvements affect water consumption, the input meter should be logged at the start of each event or change to the practice, to quantify the cost savings.

Water saving devices and other changes may require capital investment, and such cost benefits should be included in the report. Installation will take place at a later stage of the audit, or perhaps as a separate contract.

Follow-up

This last stage of the audit procedure involves routine monitoring, after a period of improvement, when recommendations have been put in place. Input meters and sub-meters should be read regularly and the site inspected to ensure that good practice is being maintained. This is best progressed via a 6-monthly or annual maintenance contract. Accurate measurement of water supply (and in some cases usage) is essential to any demand management programme. The supply side must be regularly monitored in order that so that the levels of saving achieved are known and also so that any unusual or unaccounted for increase in demand can be identified.

REFERENCES

1   WSA/WCA Engineering and Operations Committee (1994) *Managing Leakage: UK Water Industry Managing Leakage* Reports A–J: Report A – *Summary Report*; Report B – *Reporting Comparative Leakage Performance*; Report C – *Setting Economic Leakage Targets*; Report D – *Estimating Unmeasured Water Delivered*; Report E – *Interpreting Measured Night Flows*; Report F - *Using Night Flow Data*; Report G – *Managing Water Pressure*; Report H – *Dealing With Customers' Leakage*; Report J – *Techniques, Technology and Training*. London: WRc/WSA/WCA.

Appendix E

Example training programmes

AWARENESS SEMINARS (1 day)

A typical programme will include an overview of:

- Principles and concepts of leakage management (DMAs and night flow monitoring)
- Water balance components and calculation
- Economic principles (leakage targets, optimum levels, alternative strategies)
- Practical application of these principles to the company

TRAINING WORKSHOPS (5 days each)

Specific tasks for managing the strategy will be encompassed in the topic areas covered by the workshop modules, as follows:

Module 1 (1 day)

- Principles and concepts of leakage management
- Choosing a strategy (economic analysis)
- Optimum leakage levels and targets
- Understanding the components of water balance

© 2003 IWA Publishing. *Losses in Water Distribution Networks* by Malcolm Farley and Stuart Trow. ISBN: 1 900222 11 6

Module 2 (1 day)

- Advantages of continual night flow monitoring
- Design of district meter areas (DMAs)
- Design of pressure management schemes

Module 3 (1 day)

- Flow measurement and data capture
- Data analysis and interpretation
- Prioritisation of leak detection activities

Module 4 (2 days)

- Principles of leak detection; techniques and equipment
- Practical training in pipe location and leak detection
- Managing and operating a DMA scheme
- The requirements for operation and maintenance

PRACTICAL SKILLS TRAINING (continuous)

1. Flow monitoring

- understanding metering principles
- flow data capture (data logging)
- data analysis and interpretation (prioritising leak location work)

2. Leak detection techniques and technology

- location of pipes and fittings
- leak detection techniques and equipment (e.g. step testing)
- leak location techniques and equipment (e.g. ground microphone, correlator)

3. Operation and maintenance of a leakage management system

- keeping records of plans and data
- maintaining a DMA system
- maintenance of meters and equipment

In addition, the continuous training element will include on-job training for selected project staff with counterparts from consultants and contractors at certain stages of the study. These tasks include:

- Calculating water balance and analysing components
- Meter audit and calibration procedures
- Leakage studies
- Pressure management studies and PRV installation
- DMA design and implementation
- Network rehabilitation studies and techniques
- Pipe repair techniques.

Index